The Dynamics of Agricultural Change

The Dynamics of
Agricultural Change,

The historical experience

David B. Grigg

St. Martin's Press   New York

© David Grigg 1982

All rights reserved. For information, write:
St. Martin's Press, Inc., 175 Fifth Avenue, New York, NY 10010

Printed in Great Britain

First published in the United States of America in 1983

ISBN 0-312-22316-1

Library of Congress Cataloging in Publication Data

Grigg, David B.
   The dynamics of agricultural change.
   Includes bibliographical references and index.
   1. Agriculture—History.   2. Agricultural geography—History.   I. Title.
   S419.G84 1983   338.1'09   82-24034

ISBN 0-312-22316-1

For Jill, Susan, Catherine and Stephen with much love

And this shall be for music when no one else is near,
The fine song for singing, the rare song to hear!
That only I remember, that only you admire,
Of the broad road that stretches and the roadside fire.

R. L. STEVENSON

Contents

Tables

Figures

Acknowledgements

I am grateful to Tony Wrigley, Benny Farmer and Michael Chisholm for reading an early version of this book and offering helpful comments upon it, and also to two anonymous referees. Dr Verity Brack read the manuscript and helped eliminate error and polish style. The manuscript was typed by Mrs Penny Shamma, Miss Anita Fletcher, Mrs Joan Dunn, Mrs Carole Elliss and Miss Carolyn Wilby. The maps and diagrams were drawn by Mrs Rosemary Duncan, Miss Sheila Ottewell and Mr Paul Coles. I am most grateful to all these friends for their help.

Midsummer
Snaithing Lane, Sheffield
1981

The author and publisher are grateful to the following for permission to reproduce figures used in this book:

Cambridge University Press: p. 28; Paul Parey Verlag: p. 55; Edward Arnold: p. 57; Chatto and Windus Ltd: p. 69; Methuen: p. 83; Johns Hopkins University Press: pp. 114, 116; Royal Anthropological Institute of Great Britain and Northern Ireland: p. 160; Agricultural Economics Society: p. 162.

1 Introduction

Agriculture has formed the basis of economic life for over 10,000 years, and until 100 years ago the great majority of the world's population were directly employed in farming and pastoral activities. Industrialization, beginning in Western Europe and spreading slowly to European settled areas overseas, and later to many other parts of the world, has greatly reduced the importance of agriculture in the economy. In Britain and the United States only 2 per cent of the work force are employed in agriculture, and in the world as a whole less than half those gainfully employed now work in farming.

But the history of agriculture was the history of mankind until the nineteenth century, although it has not received the attention it is perhaps due. Historians have concentrated upon political history, not upon the mundane and unexciting events of farming life: even economic historians have shown more interest in the life of towns and the fortunes of industry than the ways of farmers. But although there are few agricultural historians this has not prevented them producing a voluminous and distinguished literature, which is perhaps inevitably concerned primarily with Europe and the European settlements overseas. It is for this reason that this book deals almost exclusively with Europe and North America; it is not that the rest of the world has no agricultural history of interest, but that for the most part it has yet to be written.

Those concerned with agricultural history – whether they be historians, economists, geographers or sociologists – have for the most part followed the example of historians and written narrative history, concerned with the chronology of agricultural change in one particular area, although the size of that area has varied from an English parish to the whole of Europe, and the time from a few years to a millennium or more. In the past twenty years there have been marked changes in the methods and aims of the academic disciplines which have traditionally concerned themselves with agricultural change. In the late 1950s, for example, geographers began to tire of the description of regions and the concern for unique facts about the earth's surface. They sought instead to make geography a 'law giving' science, seeking generalizations that would explain similarities rather than emphasizing the uniqueness of places. To do this they had to borrow a methodology from the more sophisticated social sciences – particularly economics – and to use statistical methods to test hypotheses and formulate models.

Economic history also experienced a sea change, beginning somewhat later, but following much the same course. In Britain, departments of economic history had grown up separately from departments of economics, and were on the whole closer bound to history. But in the last fifteen years a new school of economic history has emerged; these writers have argued that economic history can only be explained satisfactorily by the use of economic theory; that a rigorous methodology is necessary for collecting data, formulating hypotheses, and testing them. The New Economic History had its origins in the United States, but has spread to Britain. But in both countries its primary concern, with a few notable exceptions, has not been with agriculture.

Agricultural historians and historical geographers have perhaps been slow to respond to these changes in their disciplines; historical geographers however have constantly admonished each other to adopt the new approaches. Some historical geographers have had the confidence to urge agricultural historians to adopt the models of human geography and apply them to the study of agricultural change. It might be thought from reading such essays that agricultural history lacked any attempts to generalize about the nature of agricultural change, and had concerned itself only with the microscopic examination of the unique. This hardly seems fair to those who have dealt with agricultural history in the past. Are von Thünen, Marx and Meitzen to be dismissed? Have not modern agricultural historians such as Wilhelm Abel and B. H. Slicher van Bath attempted interpretations of European agricultural history on the heroic scale? There certainly have been attempts to generalize about agricultural change, although these have not always been clothed in the language and methodology of modern social science. The term model is thus interpreted liberally – indeed perhaps even in a libertine manner – in this book. What it tries to do is review for the student of agricultural history those attempts to describe and explain agricultural change which are not specific to a limited area or a particular time. In a sense this book is a systematic historical geography of agriculture. Several categories of writer are discussed. First there are those who have put forward explicit models of agrarian change – such as Karl Marx and B. H. Slicher van Bath. Second, the models of other writers can be fairly easily adapted to the study of agricultural change – Malthus, Ricardo and von Thünen are good examples – and the implications of their work are considered. Third, some models used in modern human geography may be of help in understanding the agricultural past: the diffusion of agricultural innovations is one such case. Fourth, some of the methodological approaches to the study of modern agricultural change may be usefully applied to the past. The measurement of the growth of productivity change is one such instance.

These approaches to agricultural change seem to be worthwhile, and it is the purpose of this book to review such models and their implications. The book is arranged in four parts. The first part shows how some historians and economists have seen population change as a cause of overpopulation in rural societies, giving rise to adverse changes in farm structure, land use and technology (Chapter 2). In contrast others believe population growth to be an essential stimulus to improvement in agriculture: Ester Boserup's work is the best example of this approach (Chapter 3). In the second part of the book some attempts to relate environmental and agricultural change are considered. Change in agriculture does not proceed at the same pace or in the same way in all areas. Thus Chapter 4 attempts to show how Ricardo's theory of rent may be helpful in analysing regional rates of change. Chapter 5 takes some of the ideas of modern ecologists, who deal with the flows of energy and nutrients through agro-ecosystems, and demonstrates how these approaches may help the understanding of agricultural history. Climate has often been suggested as a possible cause of agricultural change, and this presumption is examined in Chapter 6.

The great break in agricultural history came in the eighteenth and nineteenth centuries, with the industrial revolution and its implications for agriculture (Part Three). Some argue that this event transformed peasants into modern farmers, so in Chapter 7 some theories about the nature of peasant societies are examined; the following chapters consider several different aspects of industrialization and their consequences for agriculture. Structural transformation, new demands and new technologies are considered in Chapters 8 and 9, while the results of the transport revolution are dealt with in Chapter 10.

Part Four of the book deals with the idea of an agricultural revolution. Although historians have seen change in agriculture as slow, they have identified some periods of more rapid change, and dubbed these 'agricultural revolutions'. These can be usefully reinterpreted by considering the diffusion of innovations, and by applying some methods of defining and measuring productivity change. These approaches are considered in Chapters 11, 12 and 13.

Population, environment and industrialization are not the only keys to agricultural change. Many nineteenth-century writers thought institutions to be the heart of not only agricultural but all historical change, although they differed as to what was the critical institution (Part Five). The ownership of land was held by some to be of paramount importance: Marx's interpretation of agrarian change in England deals essentially with the changes in the ownership of land, and the role of enclosure in altering this, and is discussed in Chapter 14. In North America nineteenth-century historians believed that American institutions were derived from Europe. F. J. Turner in contrast thought the American environment transformed European institutions; the medium of change was the frontier, a concept which he used to interpret agricultural as well as political history.

This by no means exhausts the models of agricultural change that have been put forward, but it is hoped that it suggests to students of the economic history and historical geography of agriculture a number of stimulating ways of interpreting – and reinterpreting – agricultural history.

Part One
Population and agricultural change

2 The adverse consequences of population growth

The relationship between population growth and material wealth has long been a matter of debate. Interest has focussed upon two aspects of the problem. First, what are the relationships between population growth and the food supply? Second, how does population growth influence economic growth? There is a large literature upon both these topics, stretching back to the eighteenth century; yet comparatively little attention has been paid to the effect of population growth upon agriculture – upon land use, farming methods, productivity and farm structure. If there are few theories linking population and agriculture, nonetheless most writers have taken distinctive attitudes to the problem. First are those who believe that rapid population growth invariably has adverse effects upon the agricultural economy, causing the subdivision and fragmentation of farms, underemployment and unemployment, falling real wages, an increase in arable land at the expense of grazing land, falling crop yields, and a decline in the number of livestock that can be kept.[1]* Second are those – a minority of writers – who believe that only population growth spurs economic development and that it is the major cause of agricultural change in peasant societies, enforcing adaptations in land use intensity, and changes in the implements used.[2] In this chapter the idea of overpopulation in rural societies is discussed, and in Chapter 3 some of the positive responses to population growth are considered.

Malthus

The view that population growth can only have adverse consequences for society is closely associated with Thomas Malthus. Although Malthus did not deal with the effect of population growth on land use or on the structure of agrarian society, but on

*Superior figures refer to the Notes and references on pages 230–55.

the food supply, much subsequent work derives from his theory and it must be outlined here. He first put forward his views in *An Essay on the Principle of Population*, published in 1798, and then greatly extended this in a revised edition published in 1803.[3] He argued that population had the *capacity* to increase at a geometric progression whereas food output only had the capacity to grow at an arithmetic progression. But as population did not increase at a geometric progression there must exist checks to population increase. He recognized two categories: first, preventive checks, of which he considered only one, moral restraint or the deferment of marriage; and second, the positive checks, which included all the causes of a shortened life, but principally war, famine and disease.

What later writers have called a Malthusian cycle took the following course. Suppose population is stationary, birth and death rates being equal and income per caput at subsistence level. For some reason income per caput rises. This allows people to marry earlier and the birth rate rises; they can also buy extra food, are more resistant to disease and so the death rate falls. Thus population begins to increase. After a time, and because of the law of diminishing returns, the increase in the food supply falls behind the increase in the population. Consequently income per caput falls and as it does marriage is once again deferred and fertility declines, and as income per caput falls towards the subsistence level the death rate rises. Population growth ceases as fertility and mortality reach the same level, and income per head returns to the subsistence level. Malthus' essential conclusion was that while population can increase, it will always eventually be halted by rising mortality as income per caput falls, and that periods of temporary affluence will always be eroded by population growth, income per caput always returning to the subsistence level.

Malthus' theory was much criticized by his contemporaries and by subsequent writers. In particular he ignored the possibility of technological advance in agriculture allowing a greater increase in food output. Nor did he allow that there might be birth control within marriage, or that men might defer marriage or have fewer children within marriage in order to achieve a better standard of living. Subsequent interpreters of his theory have been apt to emphasize that population growth is halted by rising mortality rather than falling fertility, which Malthus, however, did allow as a possibility.

Optimum theory

For much of the nineteenth century food output and income per caput rose faster than population, which itself was increasing rapidly. Contemporaries believed that technological progress and investment, neither considered by Malthus, had invalidated his theory. Towards the end of the century there was renewed concern about population growth, but economists then applied themselves not to the inevitability of overpopulation but to a consideration of what was the best, or *optimum*, population for a country to have. This was done by applying the newly discovered principles of marginal analysis to the relationships between output and population.[4]

The optimum population with given resources, technology and capital is JK which maximizes output per head (Figure 1). With any population less than JK average output per head is less than the maximum, and an increase in population will give an increase in total output, average output and, in some cases, marginal output. This is due to the existence of increasing returns to extra labour inputs; the existence of such a condition has not been widely investigated but presumably occurs in sparsely populated, recently settled areas, where extra population would allow improvements

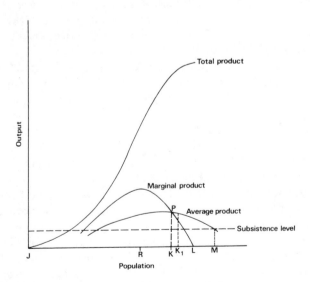

Figure 1   *The definition of an optimum population*

in transport, the spread of general overhead costs among a greater population, and the use of part of the population to clear new land for cultivation or to improve irrigation. One modern writer has argued that parts of Africa are underpopulated rather than overpopulated.[5]

At population JK marginal product equals average output per head and thereafter with a greater population output per head declines and marginal product falls below average output. At any population above JK diminishing returns are operating, and although total output increases, output per head falls, and in terms of optimum theory the country is overpopulated. However at higher populations than JK two special subcases of overpopulation may be recognized. When population reaches JL the marginal product is zero; with any subsequent addition to the work force each extra worker produces a negative marginal product. Further, between population JK and JL any reduction in population would produce an increase in output per head; modern economists have described those whose marginal product is zero as the surplus population, a measure of underemployment; there have been many attempts to measure the degree of underemployment in contemporary Asian countries.[6]

When marginal productivity reaches zero, any further increase in population leads to a *fall* in total output as marginal productivity becomes negative: eventually average output per head falls to the subsistence level at JM. Any further increase in population will lead to a rise in mortality. This point is clearly similar to the definition of overpopulation which can be derived from Malthus' theory, where population growth is halted by rising mortality, and an equilibrium position is reached where fertility and mortality rates are equal.

Optimum population then allows us to derive three definitions of overpopulation. In general terms any country is overpopulated which has a population greater than the optimum and thus an average output per head lower than the maximum possible. Two special cases can be identified. Beyond JK there is underemployment, and removal of the surplus population would lead to an increase in output per head: at JL marginal product becomes zero and not only does average output per head fall, but marginal productivity becomes negative and the total output declines. At JM average output per head falls below subsistence level, and mortality rises until output per head has returned above the subsistence level.

Optimum theory has been much criticized on the grounds that it deals only with a static situation; as population increases beyond the optimum extra resources may be employed or a new technology adopted so that diminishing returns are offset and average output per head is increased. But perhaps the major criticism is that the optimum population of an actual country has never been satisfactorily measured.

The measurement of rural overpopulation

There have been several attempts to measure rural overpopulation. Much of this work derives from investigations in eastern Europe in the 1930s, an area which was believed by many contemporary economists to be overpopulated. Two contrasting approaches can be identified, the consumption approach and the production approach.

The *production approach* is best exemplified by the work of W. E. Moore.[7] He calculated for each country in Europe in 1931–5 the net volume of agricultural output, and expressed this in standard crop units. He estimated an index of productivity for Europe by dividing the total standard crop units by the total male labour force, and from this he worked out the number of workers needed to produce the output of each country assuming that the average European productivity obtained. He then took French productivity as the norm, and derived a second series of surplus populations, which because French agricultural productivity was above the European mean, were greater than in the first case. In both cases countries with a surplus agricultural population were confined to eastern and southern Europe.

The second method of defining rural overpopulation may be called the *consumption approach*. Here an attempt is made to define the number of people that can be supported in a given area at a given standard of living. A good example of this approach is the work of Sen Gupta,[8] who calculated an index of population pressure for 335 districts in India in 1961. This index was

$$I = \frac{P - P_1}{A}$$

where I is the population pressure per km^2, P_1 is the number of rural inhabitants who could be supported at a given income per head, P is the actual rural population, and A is the total area in km^2.

P_1 was determined by $P_1 = \frac{x}{k}$ where x was the gross value of agricultural production in the district while k was the defined income. Sen Gupta took k to be the average earnings for those engaged in primary activities plus an amount 'needed to obtain a slight improvement in the standard of living'. From these calculations a relative measure of overpopulation and underpopulation in India was obtained.

Neither the production nor the consumption approach compared the actual population with the optimum population; attempts to define overpopulation in terms of consumption have been apt to compare actual consumption levels with the subsistence or near subsistence level rather than an optimum consumption level. The surplus population has thus been far less than if the comparison were made with the maximum average output per head.

Since 1945 many economists have attempted to define the degree of rural underemployment in developing countries. This has been defined as the number of the rural population whose marginal product is zero. It was thought in the past that this part of the rural population could be withdrawn from agriculture without any fall in output, and would form the basis of an industrial labour force. In the event the rapid natural increase in the urban population of the developing world has rendered this migration unnecessary. Two types of surplus population were recognized by P. N. Rosenstein-Rodan. In the first type the number surplus to requirements with the *existing* agricultural technology is calculated. A second or dynamic type is defined as the number surplus to requirements if an *improved* technology were to be introduced into the area.[9] This is clearly similar to the production approach pursued by W. E. Moore when he compared the labour available in each country with the labour which would be needed if French agricultural technology was in use.

Carrying capacity

There have been several attempts made by agronomists, geographers and anthropologists to measure the carrying capacity of parts of the tropics where shifting agriculture or bush fallowing is practised. Carrying capacity is defined as the maximum number that an area can sustain at a given subsistence level without the onset of land degradation. W. Allan,[10] one of the earliest workers on this topic, assumed that the inherent physical characteristics of the land determined the length of fallow. He then determined by survey methods: the acreage of crops necessary to maintain a family; the proportion of the area that can be cultivated; the amount of fallow necessary to allow the regeneration of vegetation and hence soil fertility. From these data he then calculated the area of crops and fallow necessary for a given soil type, and from this the carrying capacity as a density. This could then be compared with the actual density. Similar calculations have been made for parts of Brazil and the New Guinea Highlands.

Since these early works there have been criticisms of the method and attempts to expand the concept in order to incorporate differing concepts of subsistence and varying levels of labour intensity.[11]

The symptoms of overpopulation[12]

On the whole the methods of defining rural overpopulation outlined in the preceding sections have received more criticism than praise. Further, and very much to the point, they need statistical information that has rarely been available in the past. Thus many historians, and indeed those who describe contemporary events, have used as evidence of the existence of overpopulation adverse features of the land use, structure and technology which they believe are a result of population growth (Figure 2). However there is always the possibility that these diagnostic characteristics are a result of some other economic force.

The subdivision of farms

A major determinant of spatial variations in farm size in the modern world is population density, and in particular the density of the agricultural population upon the available agricultural land. If population grows and the area of farm land is not increased, or increases less rapidly than the demand for farms, farms will be subdivided. Over time the average size of farm will fall. This has happened in many parts of the developing world in the last thirty years; it seems to have happened in parts of Europe during past periods of population growth, although it must be admitted that data on farm sizes before the nineteenth century are incomplete and unsatisfactory. However differences in land tenure may influence the degree of subdivision. Thus where all the farmers are owner-occupiers the amount of subdivision will depend on whether

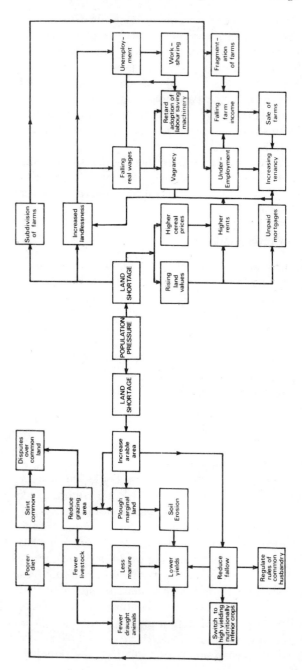

Figure 2   *Symptoms of overpopulation*
Source: D. B. Grigg, *Population Growth and Agrarian Change:*
An Historical Perspective (Cambridge, 1980), p. 45.

primogeniture or equal inheritance is practised. If primogeniture is practised and the farm passes to the eldest son, then subdivision will not occur, however rapid population growth. On the other hand if equal inheritance is practised then subdivision is inevitable. Where farms are rented from a landlord a variety of effects may be found. Thus in eighteenth- or nineteenth-century England landlords rented land to those tenants with the capital and skill to guarantee a good rent, and arranged their estate in farms whose size depended on economic and managerial principles; they felt no obligation to provide land for a growing agricultural population. On the other hand in medieval Europe landlords provided some land for their dependants, and may even have felt obliged to do so. In nineteenth-century Ireland landlords – for the most part English – did not object to the subdivision of the estates as population grew. In contemporary Latin America much land is held in very large holdings, and landlords feel no need to provide land for the rapidly growing agricultural population.

Fragmentation and field size

Where farms are subdivided as a result of population growth there is often an accompanying fragmentation. Farms which originally consisted of a contiguous block of fields are split up over a few generations and the fields of one farm intermixed with the fields of another; further, the average size of field declines. Both these characteristics have economic drawbacks, and make the introduction of new farming methods, particularly the use of machinery, difficult.

Rising land values

When the demand for farm land exceeds its supply then land values rise; both rents and the sale price of land have shown a pronounced upward trend during the periods of rapid population growth in Europe's past, perhaps most noticeably in the sixteenth century. Such a trend puts the acquisition of land beyond the reach of much of the growing rural population.

Rising agricultural product prices

In the thirteenth century, the sixteenth century and in the late eighteenth century the demand for food outran supply in Europe as population grew rapidly, and this was expressed in rising food prices. In the periods of falling or stagnant population, cereal prices either declined or were stagnant, as after 1350 and between 1650 and 1750. This connection between population growth and prices was only broken after 1850 when it became possible for Europe to import food supplies from overseas in exchange for manufactured goods.

An increasingly landless population

One of the consequences of population growth with a fixed supply of land is an increase in the number of landless, who have little alternative except to seek employment as labourers on other people's farms. Such an increase will be most marked where for tenurial reasons there is no subdivision of farms, but it can also occur where the growth of the population outruns the rate of subdivision.

Underemployment and falling real wages

Where new farms are not being created – either by subdivision or by bringing new land into cultivation – the absolute number of landless increases and the percentage of the rural population who are landless also increases. Over long periods of population growth the number of landless seeking work exceeds the demand for labour on farms and this leads to both unemployment and falling real wages. In both the sixteenth and late eighteenth centuries food prices rose more rapidly than agricultural wages. On family farms which are progressively subdivided there are more labour hours available to work than work to do, so there is under-employment. It is this, together with the unemployed landless labourers, which forms the surplus labour which many economists have tried to measure in the contemporary developing world (see pp. 23 – 6).

Retardation of technology

When there is underemployment of family labour, and unemployment of hired labour, there may be little incentive to adopt labour-saving implements and machinery. Such a condition is found in many parts of the developing world today, and existed in much of Europe before the nineteenth century. Thus the adoption of the reaper did not come until the rural population had begun a continuous decline after 1850. Nor was the scythe, long used for mowing grass, adapted to the reaping of cereals until there were seasonal labour shortages in the first half of the nineteenth century.

Land use changes

Pre-industrial European farming communities ideally required a balance of resources to satisfy their needs. Arable land provided cereals and vegetables for their food; heavy, well-watered bottom lands provided grazing and hay, but much of the grazing for livestock came from unimproved land held in common outside the cropland, which provided fodder for draught animals, sheep, and beef and milk cattle. Woodland was necessary for timber, fuel and for feeding pigs. Such a balance could be preserved when the population was small and not growing. But once population grew rapidly and densities were higher, arable land had to be expanded at the expense of grazing land. If livestock numbers were not reduced the rough grazing was overgrazed; if they were reduced then the supply of manure for the crops was less, and yields may have fallen. This vicious spiral in the European agrarian economy persisted in times of population growth until the growth of fodder root crops and clover upon the fallow.

An inferior diet

The decline in the number of livestock reduced the amount of milk and meat the peasant could produce, and cereals became an increasing proportion of his diet. In some cases population pressure enforced a switch from cereals to higher yielding crops such as potatoes, which had hitherto been grown only as fodder crops.

The use of marginal land

In some periods of population pressure, noticeably in the thirteenth century, cultivation expanded onto physically marginal land, which may have led to soil exhaustion and declining crop yields. Such a phenomenon has been noticed in many parts of Africa in the last thirty years.

Demographic interpretations of European rural history

Before the nineteenth century there are few reliable figures on national population totals, and even less upon such demographic characteristics as birth or death rates. Nonetheless there is some agreement that the period between 1100 and 1300 was one of population growth, the period 1350 to 1450 one of decline or stagnation, followed by renewed growth until the late sixteenth century. In the seventeenth century there was actual decline in many parts of southern Europe; but northern Europe – England, the Netherlands and much of Scandinavia – experienced continued but slow increase followed by stagnation from 1650 until the 1740s. Thereafter most of Western Europe had rapid increase until the early twentieth century, Ireland and France being the exceptions. In the nineteenth century migration from the country to the new industrial towns reduced the rate of rural population increase so that the agricultural populations probably did not increase much faster than they had in the sixteenth century or possibly than they had in the twelfth and thirteenth centuries. In the later nineteenth century the rural populations ceased to grow, and in many regions outmigration began to exceed natural increase and there was actual decline in the rural populations of many parts of Western Europe.

A Japanese economist, R. Minami, has argued that Malthus' assumptions about population require his cycle to follow a logistic curve, with rapid increase in the early stages being followed by a slowing rate, and finally stagnation, so that the growth of population follows a flattened S-shaped curve.[13] Three such cycles can be seen in European rural history, from 1100 to 1450, from 1450 to 1750 and from 1750 until 1900. Indeed Rondo Cameron has recently suggested that the first two cycles, covering pre-industrial Europe, form a better 'periodization' of history than the more conventional political divisions.[14] Not surprisingly some writers have suggested that European agrarian history may be usefully interpreted within this population framework. Indeed the two major histories of European agriculture, by Abel and Slicher van Bath have interpreted agricultural changes in the context of these fundamental trends, and in particular in the way population trends have influenced agricultural prices and farm labourers' real wages. Emmanuel Le Roy Ladurie has discussed the close interrelationships between population change and agrarian change in Languedoc from the fourteenth century to the eighteenth century, while Jan de Vries' account of Dutch agrarian change in the sixteenth and seventeenth centuries puts much emphasis upon demographic factors.[15]

There is little dispute that the twelfth and thirteenth centuries were periods of agricultural and population expansion. Nor is there any doubt that the fourteenth century was one of economic crisis and, after 1350, declining population. However historians differ on the causes of this crisis. For Marxist historians (see Chapter 14) the feudal system contained the seeds of its own decay; it caused inefficiency and injustice, and internal conflict between classes led to decline. Other writers put great stress upon the role of the Black Death, the bubonic plague which between 1348 and 1351 may have killed one third of Western Europe's population, while later more limited outbreaks caused continued decline in numbers until the middle of the fifteenth century. Some historians, of whom M. M. Postan is the most notable, believe that population growth slowed in the late thirteenth century due to rising mortality and that this was due to overpopulation.[16]

England in the fourteenth century

A number of writers have argued that England was overpopulated in the early fourteenth century, and that this can be recognized not only by the occurrence of the symptoms of population pressure noted earlier, but also by a rise in mortality which occurred before the Black Death, and led to a decline in total population.[17]

The only sources for the population of England and Wales in the Middle Ages are the Domesday Book of 1086 and the Poll Tax of 1377. Figures for the fourteenth-century population before the Black Death of 1345–52 can only be arrived at by deducting from the known 1377 figure an estimate of the proportion who died in the plague. This and many other difficulties of interpretation have given a wide range of estimates of total population for 1086 and *c.* 1340, from 1.01 million to 2.85 million for 1086, and between 3.65 million and 5.56 million for *c.* 1340. A consensus view suggests that the population approximately tripled, that the rate of increase was greatest in the twelfth and early thirteenth centuries, and that this rate declined in the later thirteenth century. Some historians believe that in 1086 England was already a long settled country with most modern villages already in existence, and most of the good land in cultivation. The population growth of the next two centuries led to an expansion of the area under cultivation, into forest areas in the midlands, into upland areas, led to the reclamation of marshland areas, and onto chalk downs. But this expansion was insufficient to keep up with population growth and the arable area per caput declined. Much of the land reclaimed was of poor quality, where average yields were low. Further, crop yields fell on the old established arable land; as cropping encroached onto the surrounding common land, the amount of grazing declined; thus fewer livestock could be supported, the supply of manure was reduced and so crop yields fell.

Although not all historians accept that yields fell in the late thirteenth century, or that mortality rose in the early fourteenth century, there were nonetheless symptoms of population pressure. The number of occupiers of small holdings increased, and many of these holdings were less than the 4–5 ha which it is estimated was necessary to support a family. Although rents and dues were fixed, other seigniorial impositions increased in the thirteenth century. The prices of agricultural products rose as demand outran supply, and for the increasing number with so little land that they had to rely upon day-labouring, real wages fell. In some villages the number of livestock declined, and thus probably reduced the animal content of the peasant's diet.

By no means all historians accept that England was overpopulated in the late thirteenth century. Some would argue that improvements in farming methods, particularly improvements to the plough and the cultivation of legumes such as peas and beans, and the reduction of the area in fallow, were sufficient to increase agricultural output in step with population growth. Some indeed believe England was a prosperous country on the eve of the Black Death. Other historians accept that there is evidence of a standard of living little above the subsistence level, but believe this was due to the characteristics of the feudal system, which extracted from the peasant some 50 per cent or more of his income in the form of services, fines, rents and tithes.[18]

Ireland in the nineteenth century

The Irish famine of 1845–51 was immediately caused by potato blight which destroyed much of the potato crop on which between one third and one half of the population was dependent. But most contemporaries believe that by 1840 Ireland was already grossly overpopulated. Population had risen from just over 3 million in 1754 to just over 8 million in 1841, and the decadal rate of increase exceeded 1 per cent per annum from 1751 to 1831.[19] By 1841 the density of the agricultural population to the arable area was three and a half times that in England, two and a half times that in France, and at 213 per km^2 was comparable with densities in modern Asia. Densities well above the national average were to be found in the west. Many other symptoms of overpopulation were manifest, again particularly in the west. A quarter of the farms in Ireland were less than 2 ha, two thirds less than 6 ha. Not only was the average size of farm falling in the nineteenth century but the number of people without land was increasing; in 1841 one third of rural households were without land. There was underemployment on family farms, and unemployment among labourers. From the late eighteenth century there had been an increase in the area devoted to potatoes; from the 1820s there was a shift to the more prolific but nutritionally inferior 'lumper' varieties. Fewer households could afford to keep a cow for milk. The rapid rise of rents increased the difficulty of acquiring land, and price rises in excess of increases in wages led to a fall in real wages.[20]

There is little doubt that Ireland in the 1840s was overpopulated. But there is less agreement as to what had caused this state of affairs. Contemporary English visitors blamed the profligate fertility of the Irish or the low standards of farming. Irish writers laid the blame on the English who owned most of the land and rack-rented the Irish peasantry. Free trade between Britain and Ireland had destroyed the incipient Irish industries, and prevented the possibility of rural–urban migration.

Conclusions

Although there are several prima facie cases of rural overpopulation in Europe before the age of industrialization, and although many historians would subscribe to the belief that the low level of technology prevented any European society breaking through the Malthusian barrier prior to the eighteenth century, there are nonetheless serious objections to such an interpretation. One view is that it was the inequitable control of resources, and particularly land, which kept the mass of the rural population at a low standard of living, with their surplus production being appropriated by the owners of the land. Others would emphasize the low level of agricultural technology before the eighteenth century; as long as agricultural techniques were backward, there was insufficient surplus production to support a non-agricultural population which could develop the industries and services which have caused the rapid growth of incomes per caput since 1850. But the emphasis on the adverse consequences of population growth obscures the fact that many societies have responded to the challenges of population pressure. To these approaches we now turn.

3 The positive consequences of population growth

It is widely accepted by historians that population growth may have positive effects upon agricultural change through the price mechanism. In Europe periods of population growth have also been periods of rising agricultural prices, and this stimulated farmers to expand the cultivated area and to experiment with new farming methods (see Chapter 4). A rather different approach has been taken by others who have argued that population growth has a more direct impact upon farmers, requiring them to intensify their production. Indeed some nineteenth-century writers thought that historically all areas changed from extensive pastoralism to industrialized, intensive agriculture as a result of increases in population density.[1] More recently Ester Boserup has argued that increasing population density is the major cause of agricultural change.[2]

The evidence for her model is based largely on the behaviour of peasants in Afro-Asia, and she specifically excludes modern commercial farmers using industrial inputs from her analysis. It is also based upon a number of implicit assumptions. The first of these is that population growth is independent of the food supply, and it is thus population growth that causes changes in agriculture, rather than, as Malthus assumed, increases in the food supply that allows numbers to grow. The second is that at any given stage of agricultural intensity (see Table 1) farmers are satisified with the amount of food per head they are producing and will only work harder to produce more if forced to do so by an increase in their numbers; they prefer to maximize leisure rather than profit. The third assumption is that farmers at any given stage of development are aware of a wide range of traditional farming techniques, but they only use those that require few labour inputs. Only with an increase in population will they resort to more labour-intensive techniques in order to increase output. Fourth, by intensification she means increasing the proportion of the total agriculturally

Table 1 *The characteristics of the land use stages in Ester Boserup's model*

Type	Range of population densities (per km²)	Frequency of cropping (%)	Fallow length	Type of fallow
1 Gathering	0–4	0	–	–
2 Forest fallow	0–4	0–10	One or two crops then twenty-five years fallow	Secondary forest
3 Bush fallow	4–64	10–40	Two or more crops, eight to ten years fallow	Small trees, bush
4 Short fallow	16–64	40–80	One or two crops then one or two years fallow	Wild grasses
5 Annual cropping	64–256	80–100	One crop each year with a few months fallow	Legumes and roots
6 Multi-cropping	256–512	200–300	Two or more crops without any fallow	–

usable area that is in crops as distinct from being in fallow. An increase in the cropped area and a decline in fallow is a result of increasing population density.

The sequence of land use stages that results from population increase seems to be based mainly on the experience of farming in Africa and Asia (Table 1). Where population density is low a satisfactory output can be obtained from gathering and hunting.

Tools in use	Fertilizer	Labour needs	Productivity
–	–	Low	High
Axe, fire, digging stick	Ash	Land clearance with axe and fire. No cultivation or weeding	High
Hoe	Ash, vegetation, turfs mixed in soil	Less land clearance but some weeding	Falls as yields fall and extra weeding needed
Plough	Manure and human waste	Extra preparation of seed-bed, extra weeding and carting manure	Falls as extra cultivation, extra weeding collecting and distributing manure, care of draught animals
Plough Plough	Manure, human waste, green manure, marling and silting composts	Extra cultivation, weeding, terracing, irrigation and water control	Falls as extra cultivation and weeding, collecting and distributing manure, constructing and maintaining irrigation, terracing, water control, etc.

The second stage, and the first arable stage, is *forest fallow*, where forest is cleared with fire and simple tools, there is no cultivation of the soil, and crops are sown for two years only; the land is then abandoned, the natural vegetation regenerates and soil fertility is restored; the land is only cultivated again after twenty-five years of fallow. Labour inputs are low, output per head is high. However if population density increases then farmers will have to reduce the length of the fallow and increase the period in crops in order to produce the extra food supply. In *bush fallowing*, the fallow is shorter, and fertility is not restored before the land is used again, and so crop yields fall. To arrest this decline in yields extra labour is expended on weeding and carrying manure. Thus output per man declines. Farmers thus will only change the system from forest fallow to bush fallowing in order to produce the extra food for extra numbers. Intensification is thus a result of rising numbers. The switch to bush fallowing also requires the use of new implements. In bush fallowing the fallow is occupied by grass, not trees, and the hoe is necessary to remove the grass roots before sowing. Similarly in the transition from bush fallowing to *short fallow* the plough becomes necessary; again there is an increase in labour needs not compensated for by a proportional increase in crop yields, so that output per head falls again. In the final stages of *annual cropping* and *multi-cropping*, there is an increase not only in the tasks necessary to produce an annual crop, but in capital investment in building and maintaining irrigation systems and draining land.

Thus Mrs Boserup argues that the continuous reduction of fallow and the increased labour needed to maintain crop yields mean a fall in output per head between each stage. As the farmers are assumed to prefer leisure to extra work, the only reason why they will shift from one stage to another is to produce the extra food that increasing numbers will require.

Numerous criticisms have been made of this model. Few believe that farmers are simply leisure maximizers. Where markets for agricultural produce exist, farmers will increase labour inputs to increase yields and to maximize profits; they may also increase output to meet demands for taxation or tithe.[3] Although there is no doubt that labour inputs do increase as the fallow is reduced, it is less clear that yields will fall with each successive reduction in fallow (see pp. 73–9). The model also assumes that the new techniques which are adopted with each stage give yield increases that are unable to compensate for increased labour, and so labour productivity falls progressively. However as yields in Europe doubled between 1300 and 1800, it seems likely that in the stage from short fallow to annual cropping output per man at least remained constant. Farmers of course had responses to population growth other than simply reducing the fallow. They could expand the area in cultivation as well as reducing fallow, could specialize in cash crops, emigrate, or limit their numbers. In her most recent statement, Mrs Boserup seems to have modified her original model, recognizing that farmers may migrate into unoccupied areas rather than only reducing fallowing in their immediate territory, and that some technological advances can put off the decline of marginal returns.[4]

Population growth and European agricultural history

Mrs Boserup's original model attracted the interest of archaeologists, anthropologists and geographers dealing with simple societies in Africa and Asia. Few economists or historians have applied her ideas to European agricultural history. In her more recent book she has expanded her account of intensification in Europe. But, as she has written, 'it is often difficult or impossible to determine through historical research whether the demographic change was the cause or the effect of the changes in agricultural methods'.[5] This is particularly true of the period before AD 1000. Mrs Boserup argues that when farming came to northern Europe in the fifth and sixth millennia crops were raised by forest fallow. The low population densities and the possibilities of migration when population did increase meant that forest fallow or bush fallowing prevailed until the middle of the first millennium AD. Between AD 850 and 1350 there was an increase in population. Most historians believe that a switch to the three field system and other technical advances increased the food supply and allowed this increase in numbers. However Mrs Boserup argues that population increase was due to the decline of infectious disease, particularly of the plague. The increase in numbers enforced the adoption of the mouldboard plough, the substitution of the horse for the ox for draught, the switch from bush fallowing to the three field system or short fallow. The lack of reliable evidence on the chronology of population increase or the rate of adoption of new techniques prevents any firm conclusion being drawn upon this issue (see pp. 177–8). However it seems likely that the switch to the three field system did not get under way until the twelfth and thirteenth centuries (see p. 179), which would support her belief that population growth preceded technical change. On the other hand the change in field system was not from bush fallowing to the three field system, but from the two field system, which was already established before the population increase. It had spread into northern Europe from the Mediterranean region where it was an adaptation to aridity rather than population density. Nor did the horse replace the ox in most of Western Europe until after the fourteenth century (see p. 180).

Mrs Boserup has little to say on the period between 1350 and 1750 other than to assert that in periods of falling population – presumably 1350–1450 – farming became more extensive and that the period from 1350 to 1750 was characterized by a shortage of agricultural labour, which is at odds with most interpretations of European rural history in the sixteenth century. However the case of the Low Countries might appear to bear out her argument. During the rapid population growth of the sixteenth and early seventeenth centuries the few open fields which survived in 1500 were eliminated, and turnips and clover, which had been introduced in the fifteenth century, were widely adopted in the early seventeenth century, while convertible husbandry was practised. Labour inputs in farming increased as farmers paid tradesmen and craftsmen to undertake tasks such as marketing produce, transporting manure and repairing farm implements; farmers now specialized more in farming and the subsistence household economy which undertook a variety of general duties declined. By the 1630s the Low Countries had the most intensive farming system in Europe. But the changes were not simply an intensification of arable farming by eliminating the fallow and increasing labour inputs. Much of Holland was increasingly devoted to dairying, and new land was added for both dairying and arable farming by draining polders and inland lakes, while both the north and the south of the Low Countries became increasingly dependent upon imported grain. The growth of industry in the towns led to the production of hemp, rape, woad, flax and hops while near Amsterdam a substantial area was devoted to horticulture. The towns in turn provided manure for the intensive farming. Indeed it was the extraordinarily rapid growth of the towns which promoted commercial intensive farming rather than simply the growth of the rural population.[6]

Mrs Boserup gives more attention to affairs after 1750. She argues that this increase in population was due to a decline in infectious disease rather than a rise in the food supply, a view many would share, and that the numbers employed in agriculture increased between 1750 and 1850 (see p. 188). This increase compelled farmers to shift from the three field system to rotations without fallow but including root crops and clover: that is from short fallow to annual cropping. These changes came first in those regions with high population density and last in areas of low density. The switch to mixed farming without fallow, or annual cropping, was impossible before 1750 due to the shortage of labour, but was possible after. She believes that the agricultural revolution did not improve the food supply per caput, and rising grain prices caused a switch to the consumption of the cheaper potato. During the period between 1750 and 1850 there was little increase in labour productivity in agriculture, and it was not until after the use of industrial inputs in farming that it rose rapidly.

Some of these generalizations look less compelling when examined in detail. Thus the adoption of fodder root crops and the reduction of fallow began in England about 1650 and was well under way by 1750; yet in this period there was little increase in population. There is no doubt that the new farming required extra labour, and that the labour force increased in most of Western Europe between 1750 and 1850, but technical change did lead to some increase in labour productivity, although admittedly this was more rapid at a later date (see p. 188). And if increasing population density was a major cause of intensification why were Ireland and Norway not more progressive in the century after 1750, for here growth was rapid and densities higher than in most other parts of Europe? The adoption of the potato does however bear out her views. The potato was known in the sixteenth century, but little grown until the eighteenth century; but with the exception of Ireland, it did not become of major importance until after 1815. By the 1840s it was a major part of the diet in the most densely populated parts of Europe – Ireland, Norway, northern France, Belgium and parts of Holland.[7]

Conclusions

Mrs Boserup's assertion that population growth causes agricultural change rather than being a function of it has provoked much reinterpretation of agrarian change in simple peasant societies. In Europe matters were already more complex at an early date, for agriculture was commercialized, urban demand important, and technical advance considerable. While population growth was unquestionably a potent factor in causing agricultural change, there are too many alternative explanations of the events associated with periods of population growth for it to be the only cause.

Part Two
Environment and agricultural change

4 Price, technology and environment

Geographers have always been concerned with the impact of the physical environment upon spatial variations in agricultural activity, and quite rightly so, for environment is a powerful influence upon the distribution of agricultural land, upon the type of crop grown, livestock raised, and upon agricultural productivity. In the past much emphasis has also been put upon environment – particularly climate – as a factor causing changes in human affairs over time. Indeed many historians and geographers have argued that it is the prime cause of economic change. Of these Ellsworth Huntington is probably the best known, although he had a distinguished predecessor in the English historian H. T. Buckle. More recently A. J. Toynbee argued in his massive *A Study of History* that the progress of civilization was in part a response to the stimulus of a harsh environment, a view taken much earlier by Karl Marx, who wrote:

But it by no means follows from this that the most fruitful soil is the most fitted for the growth of the capitalist mode of production. . . . It is not the mere fertility of the soil, but the differentiation of the soil, the variety of its natural products, the changes of the seasons, which form the physical basis for the social division of labour, and which, by changes in the natural surroundings spur man on to the multiplication of his wants . . . .[1]

Geographical determinism came under considerable attack in the 1920s, and interest in environment as a factor in agricultural change declined until the 1960s, since when there has been a revival of interest in climate and history.[2] There are however good reasons why environment has not commonly been considered as a major factor in promoting agricultural change.

First, changes in the environment over time can often be shown to be a result of man's activities. Thus, for example, the silting of the lower courses of rivers in the Mediterranean basin and the creation of ideal conditions for the breeding of the *anopheles* mosquito, was a result of overgrazing in the upland catchment areas. More recently some authorities believe that human activity – in burning coal, for example, or destroying large areas of rainforest – has led to changes in climate.[3]

Second, it is difficult to show that there have been autonomous changes in the physical environment – independently of man's activities – over the last two millennia except in the case of climate and changes in sea level. The latter for example made it easier to reclaim land in the Low Countries in the twelfth and thirteenth centuries;[4] the influence of climatic change is considered in more detail later (see pp. 81–5).

Third, modern thought has emphasized man's manipulation of environment. Indeed the whole history of agriculture can well be interpreted in these terms. The Neolithic period saw the domestic-ation of crops and livestock; since then much of the advance in agriculture can be seen in terms of human alteration of the environment to farm larger areas more productively, a history that includes the control of floodwaters of great rivers, the irrigation of arid lands, the exclusion of the sea from coastal regions, the wholesale destruction of forest land to provide open land for the cultivation of cereals, the underdrainage of waterlogged clays and peats, the creation of flat land on steep slopes by terracing, and the amelioration of low temperatures by windbreaks, stoves and glass-houses. In short the history of agriculture appears to modern writers to be explicable only in terms of man's manipulation of nature and not of nature's manipulation of man.[5]

Fourth, even where it can be shown that there are environmental changes which precede or are contemporary with changes in agriculture, it is difficult to assign cause unequivocally to environ-mental change. First, there may be alternative and equally plausible explanations; thus in the eleventh and twelfth centuries settlements appear to have been created in certain upland areas for the first time. Some have attributed this to the increase in temperatures, but it is equally explicable in terms of growth of population and the subsequent rise in grain prices. Second, environmental change is often permissive in nature: it allows a particular type of change but cannot be said to cause it. Thus the great increase in the cultivation of spring barley in Britain since the 1930s may be related to a run of warmer conditions in March; but it cannot be explained without reference to the rapid growth in demand for barley for fodder and the improvements in barley breeding.[6]

For these reasons it is difficult to see – except in the case of climate – the physical environment as a primary cause of agricultural change. On the other hand it is impossible to ignore the physical environment in interpreting agricultural change. Agriculture, unlike most other economic activities, deals with living things, whose productivity is based not only upon their inherent biological characteristics but upon their interrelationships with the physical environment. In the following two chapters the idea of agriculture as an agro-ecosystem is considered, and then the impact of climatic change. In this chapter the relationships between prices, technology and environment are discussed.

Environmental theories of agricultural location

It has proved surprisingly difficult to establish an environmental theory of agricultural location, although the importance of environmental variations influencing crop and livestock distributions is readily apparent. One reason for this is the great number and complexity of the influences that determine what crop is produced at any point on the earth's surface. Not only are many socio-economic and cultural factors important, such as the relative location to market, the farmer's age and managerial abilities, the demand for different products and the availability of capital, but also a large number of environmental factors. Nor is it always clear which particular environmental factors determine the location of a particular crop, whether it be, for example, the length of the growing season or the amount of precipitation received in early summer, or the hours of sunshine in the ripening period. Contemporary analyses of crop distribution are made more difficult by the fact that the farmer can modify the environment by supplying fertilizers or irrigation water, or use early ripening varieties to extend cultivation beyond the apparent limits. The problems are compounded by the fact that not all crops have the same climatic requirements for optimum growth; the difference between temperate and tropical crops best illustrates this point, but there are also instances within smaller areas. Thus within England wheat yields are inversely related to rainfall in June, July and August, while grass yields are directly correlated with rainfall in these same months. It follows from this – and for other reasons – that it is impossible to rank land on an absolute scale from good to bad.[7] Not only does the value of agricultural land to a farmer vary with changes in the price of products, but technological changes in farming methods or transport facilities may lead to a reappraisal of land quality. However the rudiments of a theory of agricultural location may be derived from Ricardo's theory of economic rent; this may be combined with K. H. W. Klages' idea of the ecological optimum for crops to give a spatial component. From these ideas may be derived several useful interpretations of agrarian change.

Ricardo's theory of rent

Ricardo argued that each factor of production received its return: profit on capital, wages for labour, and rent for land.[8] But rent did not mean the contract rent which a tenant pays a landlord for the lease of a farm, but that 'portion of the produce of the earth which is paid to the landlord for the use of the original and indestructible powers of the soil'. It thus excludes rent paid for the improvements which have been made to the land in the form of buildings, hedges, and reclamation. Economic rent is then that part of the revenue above production costs but excluding the farmer's remuneration for his capital improvements, his labour and his managerial skills and excluding that return to the *landlord* attributable to his capital investment in the land. Thus economic rent as distinct from contract rent is difficult to measure and has mainly an expository value. Ricardo believed that economic rent arose from differences in the quality of land. Suppose a town surrounded by land of three different qualities, Grades I, II and III. All farmers have the same managerial abilities, each uses the same amount of labour and capital inputs per acre and the cost of transporting wheat to market is assumed the same for all farmers. With demand at a given level it can be supplied by farmers occupying only Grade I land, who, because they all have land of the same fertility and use the same number of inputs, each gets a yield of 2690 kg/ha. This land pays no economic rent. Demand then increases and land of Grade II quality is occupied; farmers on this land use the same number of inputs as farmers on Grade I, but because of lower fertility they get only 2018 kg/ha. Thus an economic rent of 673 kg accrues to Grade I land, the difference between yields on Grade I and Grade II land, a result of the difference in the fertility of the soil. Demand increases further, and to satisfy it Grade III land is brought into cultivation. This yields only 1345 kg/ha, and thus a rent of 673 kg arises on Grade II land, and that on Grade I increases to 1345 kg. Ricardo did not further elaborate this theory of rent due to soil fertility but some essential ideas can be inferred from his writings.

First, the extensive margin of cultivation where no economic rent is earned is due to differences in soil fertility, not the cost of transport, as in von Thünen's model (see p. 136). Second, the extensive margin of cultivation varies with changes in demand and price; as price rises so infertile land is brought into cultivation, as it falls the least fertile land is the first to go out of cultivation. Third, Ricardo recognized the possibility that instead of bringing extra land into cultivation farmers might increase inputs on the existing land; this is the intensive margin of cultivation which was elaborated by von Thünen. Fourth, Ricardo related land quality to the cost of production. Grade I land gives higher yields and lower production costs per kilogram because of the difference in soil fertility.

The ecological optimum

The idea of the ecological optimum for crop plants was introduced by an American botanist, K. H. W. Klages.[9] He argued that for any crop there are minimum requirements of moisture and temperature

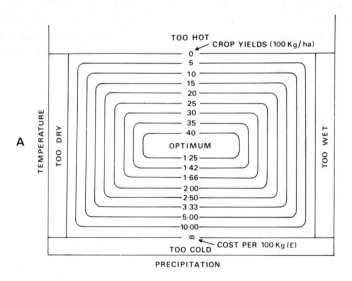

Figure 3   *The ecological optimum*

without which the crop will not grow, and also maximum conditions beyond which growth ceases. These minimum and maximum requirements vary from plant to plant, and provide a spatial limit beyond which a crop cannot be grown. There are also *optimum* conditions for any plant; there is a part of the earth's surface where physical conditions are ideal and this area gives both the highest average yield and the lowest yield variability. This idea is demonstrated in Figure 3 using two environmental variables, temperature and rainfall. The optimum area is in the centre of the region; northwards temperature rises and exceeds the maximum requirements of the crop so that yields fall; southwards temperatures fall below the optimum and yields also decline. As production costs per hectare are the same at all locations, cost per kilogram rises away from the optimum. Ultimately yield falls to zero and an absolute physical limit to cultivation is reached.

If the cost of production and the price of the product are introduced then an economic ecological optimum can be established. Figure 4 is based on the costs and prices in England in the 1960s, when wheat was £2.50 per 100 kg and total costs for producing wheat were £50/ha. Table 2 shows gross return per hectare for different yields. The distinction between the physical margin, with a zero crop yield, and the economic margin, where return per hectare equals production costs per hectare, occurs when the yield is 2000 kg/ha. As can be seen in Figure 5, where the same data are plotted in a different way, there is a plateau area of net return of £50/ha, with a fall to the economic margin where production costs equal gross return. If wheat is grown with yields less than 2000 kg, there is a negative net return.

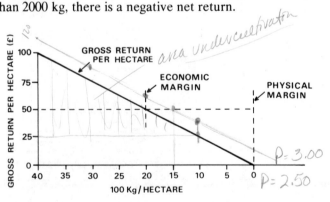

Figure 4   *The economic and physical margins*

Table 2 *Costs and returns of wheat at different yields*

Yield of wheat (100 kg/ha)	Price per 100 kg		Production cost (£/ha)	Gross return (£/ha)	Net return (£/ha)	Cost per 100 kg produced (£)
40	2.50	3 .00	50	100	50	1.25
35	2.50		50	87.5	37.5	1.42
30	2.50		50	75.0	25.0	1.66
25	2.50		50	62.5	12.5	2.00
20	2.50		50	50.0	0.0	2.50
15	2.50		50	37.5	−12.5	3.33
10	2.50		50	25.0	−25.0	5.00
5	2.50		50	12.5	−37.5	10.00
0	2.50		50	0.0	−50.0	∞

This analysis thus distinguishes between the economic and physical margins of the cultivation of a crop. It also demonstrates how the area under cultivation can change with changes in either the price of the product or the cost of production. Thus if prices rise and the cost of production remains constant a larger area will produce a positive net return; conversely a fall in prices will lead to a smaller area giving a net return and cultivation will contract to the better land. Changes in production costs without changes in prices can have similar effects, a fall in production costs leading to a larger area within the economic margin, a rise reducing the area where production is viable.

Agricultural prices and environment

Most modern analyses assume that farmers aim to maximize profit; hence farmers select those products where the spread between price and production cost is greatest, or where the cost of the marginal input equals the price of the marginal output. But not all farmers are profit maximizers (see Chapter 7) and not all farmers use the most cost-effective production methods (see Chapter 11). Further, the farmer's response to changes in price is not always what might be expected. For a variety of reasons responses to changes in prices are slow and dramatic changes in land use are rare. The view of one Lincolnshire farmer in the late eighteenth century may stand for the view of many, then as perhaps now:

...as farmers are accustomed to see great and sudden fluctuations in the

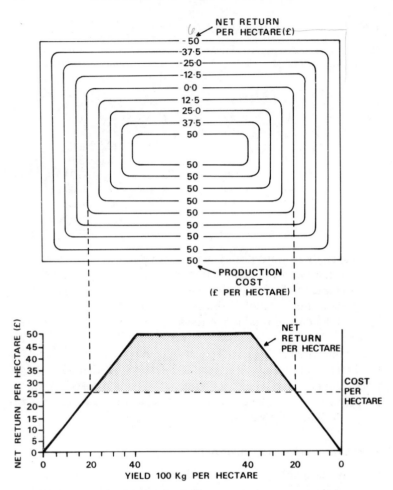

Figure 5 *Optimum area and the margin*

marketable value of the different products of their industry, without much affecting their average value from year to year, such a crisis in rural affairs must manifest itself by effects sufficiently steady, continued and uniform before this experienced and wary class of men will be persuaded to alter materially their general course of management.[10]

The long term trend in cereal prices in Europe is known (Figure 6). The general trend was upward in the thirteenth century and then slowly downward in the later fourteenth century; the sixteenth century saw a remarkable rise which continued in most countries until the mid seventeenth century; from then to the mid eighteenth century wheat prices were either falling or stagnant. Thereafter wheat prices again rose until the mid nineteenth century – except in England where the decline came earlier – and fell in the later nineteenth century, recovering by the early twentieth century. These major phases have been related to periods of expanding population, when demand outran supply, and stagnant or declining population, when demand was more easily matched by supply and prices were lower. In the later nineteenth century the fall in prices was primarily due to the import of cheaper grain from North America and from other overseas suppliers. There are no long run series of the prices of livestock products comparable to those for cereals, but it is generally believed that they followed the general trend of cereal prices, with the important proviso that livestock prices rose less than cereal prices during the periods of boom, and fell less in the periods of decline, a trend which had important consequences for farmers' choice of enterprise.[11] Although price trends are an important measure of agricultural change, they need to be complemented with long run trends in the cost of inputs, for

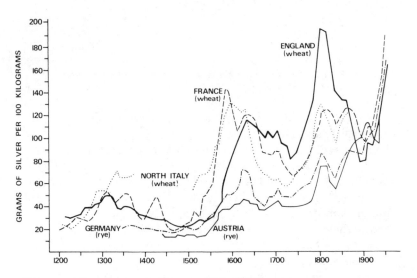

Figure 6   *Cereal prices in Europe, 1200–1950*
           Source: W. Abel, *Agricultural Fluctuations in Europe: From the Thirteenth to the Twentieth Centuries* (London, 1980), p. 2.

it is the relationship between the two that determines the profitability of farming. Unfortunately there are no long run data on the cost of farm inputs, although it is likely that labourers' wages followed the trend in cereal prices, but with a considerable lag, so that labour costs declined in real terms during periods of rising prices, and rose in real terms in periods of falling or stagnant grain prices.[12]

Ricardian theory would lead us to suppose that the area in crops would increase in periods of rising cereal prices, and decline in periods of falling price. Further, during periods of rising price land of poorer quality would be brought into cultivation, while in periods of falling prices the poorer quality land would be the first to be abandoned. B. H. Slicher van Bath has made the fullest inquiry into these effects. Dividing European history into periods based upon price movements (Table 3) he has argued that there were characteristic responses in each boom and slump.[13]

Table 3   *Price changes in Europe, 1150–1850*

Period	Character
1150–1300	Agricultural boom
1300–1450	Severe agricultural depression
1450–1550	Slight agricultural recovery
1550–1650	Agricultural boom
1650–1750	Slight agricultural depression
1750–1850	Agricultural boom

Source: B. H. Slicher van Bath, *The Agrarian History of Western Europe AD 500–1850* (London, 1963), p. 113.

In boom periods there was an expansion of cultivation into marginal areas: farmers had the money and the confidence to invest in reclamation, new buildings and to experiment with new farming methods. Not only was cultivation extended into hitherto uncultivated land, but rising cereal prices sometimes encouraged the ploughing of permanent grassland. The higher prices for farm products made it possible to support a family on a smaller area, and so there was an increase in small holdings (but see pp. 28–9).

In periods of falling cereal prices there was initially little reaction from farmers, many of whose costs are fixed; indeed in an attempt to compensate for lower prices, output was often increased, and this oversupply led to a further decline in prices. Then there were changes. Poorer land was abandoned, and because there was a smaller decline in livestock product prices than cereal prices, there was a shift towards livestock production, and in some areas arable land was converted to permanent pasture. In areas near to towns some farmers turned to industrial crops such as dye plants or flax, and there was a growth of domestic industry in rural areas. Poor prices inhibited investment and thus agricultural progress was slow in these periods. However crop yields did not necessarily fall, for it was the poorer land that was abandoned, while the increase in livestock farming may have increased the supply of farmyard manure.

Although most historians would support Slicher van Bath's account of the trend of the expansion and contraction of arable farming in response to changes in cereal prices, most of the evidence on which this analysis is based is not derived from

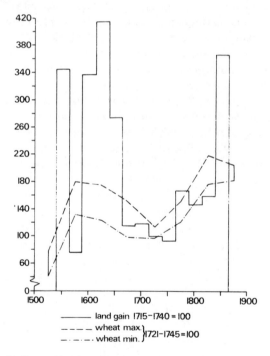

Figure 7   *Indices of land gained in the Netherlands by polders compared to wheat price indices, 1525–1875*
Source: B. H. Slicher van Bath, *The Agrarian History of Western Europe AD 500–1850* (London, 1963), p. 201.

national statistics, which did not exist until the mid nineteenth century, but from an examination of the literary evidence for small areas and for short periods. One of the few reliable estimates of the increase in arable land comes from the Netherlands. The increase in the area embanked and drained between 1525 and 1875 shows a close relationship with trends in cereal prices (Figure 7). The ebb and flow of the cereal area over a much longer period has been traced by pollen analysis at a site in Germany. The proportion of pollen from cereals in the soil profile follows the rise and fall of grain prices remarkably closely.

There seems widespread agreement that the area under cultivation expanded in the twelfth and thirteenth centuries. Much of this was by movement into forest areas and commons that surrounded existing village lands. But there was also movement into marginal areas; first into upland areas. In all temperate regions the fall of temperature with altitude and the increase in rainfall reduces the growing season and often gives acid soils and waterlogging, so that yields are lower and yield variability higher. Upland areas are thus only used for crop cultivation in periods of very favourable prices. Thus in the twelfth century there was an upward movement in Alpine valleys, in the southern uplands of Scotland, and on chalk downland in southern England. A second distinctive movement in the twelfth and thirteenth centuries was into areas whose surface drainage was poor and which required considerable investment before they could be cultivated, and marshland areas on the coasts of the Netherlands, in Lincolnshire, Norfolk, and the Thames estuary which all required embankment.[14]

The revival of cereal prices in the early sixteenth century saw a similar movement into marginal areas; the late sixteenth and early seventeenth centuries saw renewed interest in fen drainage and coastal embankment in England and the Netherlands. The third period of rising prices, beginning in the 1740s, saw a further expansion onto marginal land, an expansion which is much better documented than in earlier periods. Thus in England and Wales in 1800 some 21 per cent of the total area was still unenclosed wasteland. This was overwhelmingly in the upland areas of Cumbria, the northern Pennines, Wales and Cornwall; in the lowland counties there was little and it was mainly in the fens and on poor sandy soils such as the Brecklands. Although wheat prices fell after the end of the Napoleonic Wars, wheat prices were still high compared to the pre-Napoleonic period, and reclamation and enclosure continued until the fall of wheat prices in the 1870s. By 1873 only 6 per cent of England and Wales remained in unenclosed waste.[15]

Cereal prices and the contraction of arable land

Slicher van Bath has argued that periods of falling cereal prices have seen an abandonment of the poorest land, and a switch to livestock products, whose prices generally fall less than cereal prices in periods of agricultural depression; in some cases arable land is thus converted to permanent grassland, both because of the more favourable prices but also because of savings in labour costs.

It is in fact difficult to measure the decline of arable land in depression periods, and even more difficult to establish whether this took place first upon marginal land. The existence of medieval tax data allows this to be tested in thirteenth-century England. In 1342 the *Nonarum Inquisitiones*, a tax upon the value of corn, wool and lambs, required each parish in England and Wales to return an assessment of the value of these products, and to explain any differences from a tithe on clerical incomes made in 1291. Many parishes explained a decline in value as due to the abandonment of arable land. The areas in which references to abandoned land were most numerous were the North Riding, Shropshire, Sussex and a group of counties to the north west of London. A study of these latter counties has shown that in Bedfordshire it was the light soils of the uplands that were abandoned rather than the clay vales, but in Cambridgeshire abandonment was greatest on the upland clay areas rather than on the chalk uplands or the lowland fen, while in Buckinghamshire there was no correlation between soil type and abandonment.[16]

Over the period between 1350 and 1450 there was no decline in cereal prices, but wheat prices remained low until the inflation of the sixteenth century. During this former period a large number of villages were deserted in England and Wales. Many of these were on thin limestone soils in the uplands of the East Riding and Lincolnshire, or on poor sandy soils in west Norfolk; but not all abandoned villages were on marginal land. Indeed in the period between 1450 and 1520 most village desertions seem to have been due to the enclosure of arable for its conversion to grassland as wool prices moved more favourably than wheat prices. In the later sixteenth century, when the number of desertions declined, many of the abandonments took place in the Midlands on clay soils which could be used for either arable or grass, and were thus not marginal land.[17]

The sixteenth and early seventeenth centuries saw a continuous increase of cereal prices. But from 1650 to 1750 grain prices were either stagnant or declining; cereal prices in the second quarter of the eighteenth century were between a quarter and a third lower than the average of the 1660s. But this does not seem to have been followed by any abandonment of arable land; on the contrary chalk upland areas which had been largely given over to sheep grazing were converted to cropland. On the other hand the lower declines in the prices of livestock products prompted the conversion of much heavy clayland in the Midlands to grassland.[18]

The period after the 1870s saw a continuous decline in cereal prices, and especially wheat prices, with brief interruptions in 1900–18, down to the 1930s. This is the only period of English agrarian history with falling wheat prices for which data on area are available. During this period, according to the statistics, relatively little land was abandoned, although undoubtedly much enclosed upland grassland deteriorated to rough grazing. The primary response was a fall in the area under cereals, and particularly in wheat, and this was reflected in an overall decline in the area in tillage. The relative decline in tillage was greatest in the west where crops were least important, and in the Midlands where heavy clays could be converted to grass. In the east and south where cereals predominated there were greater problems, for the dry summers precluded the good growth of grass. The poor sandy soils of the Brecklands and the thin limestone soils, which required much manuring and labour, suffered most, while in the dry south east the heavy clays of Essex suffered badly from the depression. There were thus fewer alternatives for farmers and the decline in cereals was less marked and the consequences of the depression greater. Some estimates of the fall of rent in the east Midlands indicate that it was the least fertile soils that suffered most. On good soils rents fell 30.8 per cent between 1878–9 and 1893, on medium soils 38.4 per cent and on inferior soils 43.6 per cent.[19]

Technological change and land revaluation

Changes in price may well alter the use of land, either by the reclamation or abandonment of marginal arable land or by the conversion of arable land to grassland. Equally important, changes in technology may alter the possible uses of land and a farmer's valuation of a particular piece of land. Such instances abound. Much arid land has little agricultural value other than for nomadic grazing but the establishment of an irrigation system can convert poor grazing land to first class arable: the remarkable changes in the Punjab in the late nineteenth and early twentieth centuries illustrate one such case. The ability to remove excess water can have equally dramatic effects. At the beginning of the nineteenth century the interior fenlands of south Lincolnshire were regularly flooded in winter and early spring and could be used only for the summer grazing of cattle and sheep. But the introduction of steam-pumping and the lowering of the beds of the main drainage channels allowed their conversion to highly productive arable land in the early nineteenth century. In Victoria, Australia, the cost of clearing the mallee, a scrub vegetation in which *eucalypt* species were dominant, was prohibitive until the invention of the stump-jump plough and the scrub-roller in the 1870s; subsequently there was a rapid expansion of wheat growing.[20]

One interesting revaluation of land will be dealt with in a little more detail. In the mid seventeenth century the clay areas of lowland England provided the bulk of England's wheat output; these 'heavy' soils were also in places in grass for grazing cattle. In contrast the 'light' soils– limestone and sandy soils – were still largely in waste or poor grazing for sheep; cereal crops, where grown, gave low yields and the predominant crop was barley. However in the eighteenth century the new husbandry, in which turnips and clover were grown in rotation with wheat and barley, spread rapidly on the light soils of eastern and southern England; but as roots could not easily be grown on the clay soils, there was less progress, and land continued to be cultivated with a fallow. The fall in grain prices after 1815 caused great distress to the clay or heavy land farmers, but on the light land farms crop yields, stock densities and land values continued to increase. By the middle of the nineteenth century the light lands had become the major cereal producers, and progress on the clays only began when underdrainage was undertaken, or when farmers converted their land to grass and benefited from the trend towards higher livestock prices after 1850.

It has thus been argued that introduction of the new husbandry was only possible upon the light lands; and that the great increases in the farming productivity of the new system led to higher profits, rents and land values on these soils. Conversely progress on the heavy clay soils was much slower, because of their unsuitability for the new system, and they ceased to be competitive as grain producers. Their salvation was in conversion to grassland. These arguments have been applied to two periods in English agricultural history; the period 1650–1750, and that after 1815. The characteristic features of heavy and light soils are relevant to both periods.[21]

The heavy soils, mainly developed on clays, were poorly drained. The high clay content and the small pore spaces meant that the soils were impermeable and often waterlogged, while as the soils were mainly on easily eroded clay and till deposits most clay soils occurred in lowland areas with low gradients and thus surface drainage was difficult. Ridge and furrow was the only means of removing surface water, and under drainage was not practised until the early eighteenth century and was probably not very efficient until the introduction of pipes, tiles and the mole-plough in the mid nineteenth century. Heavy soils were also physically difficult to cultivate; they required three or four horses to each plough and were thus more expensive to cultivate than light soils. In autumn clay soils return to field capacity after the summer soil moisture deficit much sooner than light soils, and so have a shorter period for autumn cultivation than light soils. Thus in some wet autumns a good seed-bed cannot be prepared rapidly enough and often autumn crops cannot be sown. Excess soil moisture inhibits root development and limits the crop's access to plant nutrients. In the spring soil temperatures rise slowly as insolation is used in evaporating water in the soil, and the growing season is thus shorter than on light soils. Above all the clays were not suitable for root crops, because of the difficulties of lifting the crop in a wet autumn; if sheep were fed *in situ* soil structure was damaged and the animals were liable to foot rot. The absence of a root crop for fodder limited stock densities, the supply of manure and thus the possibilities of increasing cereal yields.

Light lands lacked many of these disadvantages. They were free-draining, and the friable soil was easily and thus cheaply cultivated. Their principal drawback was a low plant nutrient status. However when the new husbandry was adopted clover and roots gave fodder for a greatly increased number of livestock; their dung, mixed with straw to give farmyard manure, could be applied to the land and increased crop yields. Before the growth of turnips and stall-feeding, much livestock dung was wasted on the commons.

The low cost production of the light soils allowed farmers to market grain even in the periods of poor cereal prices between 1650 and 1750. In this period, in spite of uncertain prices, the new husbandry was widely adopted in the light soil areas; with lower production costs, they could undercut the clay grain producers, and make a profit. Sheep and cattle were kept, and their manure not only maintained soil fertility but they gave meat and wool. In the low clay vales however costs did not fall, and cereal producers

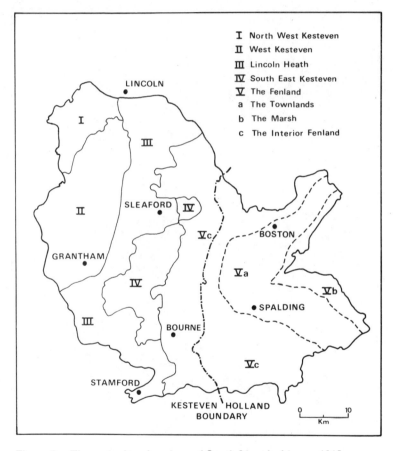

Figure 8   *The agricultural regions of South Lincolnshire, c. 1815*
Source: D. B. Grigg, 'Changing regional values during the agricultural revolution in South Lincolnshire', *Transactions and Papers of the Institute of British Geographers*, no. 30 (1962), p. 95.

were in difficulties; only in midland and western England, where grass yielded well on clays, was it possible to shift to livestock production. Thus although corn output may have fallen on the clays, this was more than compensated for by the expansion of cereal production on the light soils, while livestock production on both types of soil increased.[22]

The difference between the light and heavy soils of lowland England is again clear in the period after 1815 when wheat prices

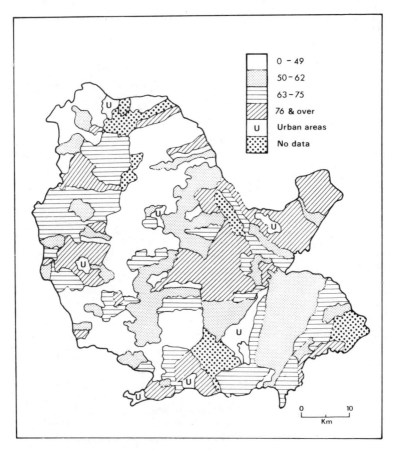

Figure 9   *Rent in 1815, shillings per hectare*
            Source: D. B. Grigg, 'Changing regional values during the agricultural revolution in South Lincolnshire', *Transactions and Papers of the Institute of British Geographers*, no. 30 (1962), p. 97.

fell dramatically from the high prices of the Napoleonic Wars, livestock prices less dramatically. For this period parochial assessments of the value of land and buildings are available which give a reasonable guide to the rent per hectare of agricultural land for three clayland regions in south Lincolnshire, the light soils of Lincoln Heath and north west Kesteven, and for the Lincolnshire fenland (Figure 8). In 1815 the lowest rents per hectare were on the central light soils, where there was still land in rough grazing and warren, and crop yields were low; rents were also low on the gravel soils of north west Kesteven (Figure 9). Rents were higher on the Lias clay vale of the west, where wheat yields were higher than on the limestone soils. In much of the undrained fenland rents were low, but the coastal marshes were better drained and high rents per hectare were to be found. Over the next thirty years enclosure and the adoption of the new husbandry transformed the heath, and the introduction of steam-pumping allowed the conversion of much of the fenland to rich arable. On the clays however there was little progress either in west or south east Kesteven and rents rose very little (Figure 10).[23]

Biological changes and land values

So far it has been shown how price changes may affect different types of farming and different environmental regions in different ways, and how technological advance may improve the quality of land. But biological improvements of plants or animals may also change the value of farmland. Foremost among such instances are improvements in plant breeding; crops may be bred to have a variety of characteristics – to respond to fertilizer, to be immune to specific diseases, or as in the case of the sugar-cane or beet, to have a higher sugar content. In this century some of the more notable advances have been in breeding cereal varieties that will germinate and ripen in a shorter growing season. These have permitted the advance of crop cultivation into areas which could not previously be occupied. This may be illustrated by the case of the Canadian prairies. This large area of flat land west of the Canadian shield was virtually unsettled until the 1870s. The area in crops was less than 500,000 ha in 1891 but over 16 million ha in 1931, while the area in wheat rose from 1 million ha in 1901 to 10 million hectares in 1931.[24] There was a wide variety of reasons for this remarkable expansion. In the United States settlement was largely complete by 1891, and settlers then moved north into Canada. The construction of railways, together with preferential freight rates, made exports of wheat to Britain possible.

The prairies did have problems however. In the west aridity presented difficulties, but elsewhere it was the shortness of the growing season. The critical factor was the possibility of frosts in summer and autumn which could cause partial or even total crop failure. The length of the effective growing season – the average

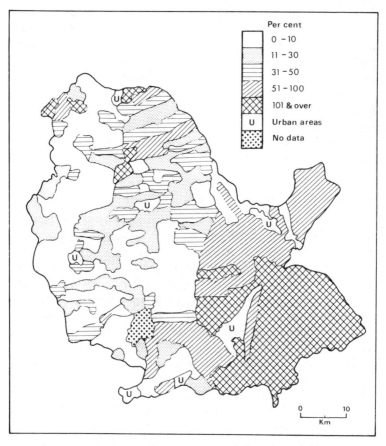

Figure 10 *Percentage change in rent per hectare, 1815 to 1840–7*
Source: D. B. Grigg, 'Changing regional values during the agricultural revolution in South Lincolnshire', *Transactions and Papers of the Institute of British Geographers*, no. 30 (1962), p. 98.

number of days between the general seeding of the wheat crop and the first summer killing frost – was 170 days in the south of Manitoba, but fell to 130 days only 200 miles to the north.[25] The initial settlement of Manitoba was made possible by the use of the hardy wheat variety Red Fife, a Galician wheat imported into Ontario in the 1840s by a Scottish immigrant, David Fife. He marketed this seed, which became popular in Minnesota and the Dakotas, and from there was taken to the prairie provinces. Not only was it early maturing but hard and with a high gluten content, making it easy to mill with the new roller process and popular with millers for its flour quality. But it was the breeding of Marquis which allowed the expansion of wheat cultivation to its modern limits in the north; it was bred within the provinces in government plant breeding stations, and was in general use by 1909–12. This allowed the first settlement of the Peace River valley, but the successful cultivation of wheat in this particularly northern region was only possible with the breeding of Garnett and Reward, bred in the 1920s.[26]

Comparable extensions of the poleward limits of wheat cultivation by the breeding of early maturing varieties have occurred in Scandinavia and the USSR. In Finland experimental work on early maturing spring wheat and hardy winter wheat began in Helsinki in 1909; by the 1940s the northern limits of winter wheat had moved northwards 240–320 km. Spring wheat, hardly grown at all in the early twentieth century, was being grown north of the Gulf of Bothnia by 1946. This northwards extension was assisted by some increase in summer temperatures in the interwar period; the subsequent decline combined with rising production costs has led to a contraction of wheat cultivation since the 1940s.[27]

Conclusions

Sufficient has been written, it is hoped, to demonstrate some of the interrelationships between technology, environment and price changes, and their significance in interpreting agrarian change in the past. Historians have of course not neglected the role of technological change in agricultural history; and the modification of the land to improve its farming potential has received adequate attention. On the other hand the regional differences that arise in periods of technological or price changes perhaps still require further exploration.

5 Agricultural systems as ecosystems

Until the 1930s most ecologists studied the members of one species, or one individual of that species and its interrelationships with the inorganic environment. But it was then argued that there was a much more complex web of interrelationships between plants, animals and the environment; this 'interaction system comprising living things together with their non-living habitat' was called an ecosystem.[1] Such a viewpoint has since become widespread in ecological studies. Three major approaches have been developed. First are the attempts to measure the biological productivity of different parts of the earth's surface. Second is the measurement of energy flows through an ecosystem, together with the investigation of food chains and trophic levels, and third is the study of the nutrient cycle.

More recently the idea of the ecosystem has been applied to the study of agricultural systems. Indeed it has been argued that 'agricultural systems are ecosystems which are modified and managed by man to produce outputs which are useful to man'.[2] There are clearly important differences between the natural ecosystem and the agro-ecosystem. The former, unlike agriculture, has no final product which is removed from the system, while agro-ecosystems also have inputs such as fertilizers which do not occur in natural ecosystems. Nonetheless the approach has proved useful in studying modern farming systems, and may have some value in interpreting changes in the past.[3]

Food chains and energy flows

All living matter derives its energy from the solar radiation received at the earth's surface. Green plants which contain chlorophyll can convert sunlight into carbohydrates by photosynthesis, thus fixing the sun's energy. Some of the carbohydrates are used simply to maintain the plant – respiration – but some are available for herbivores to eat; but there is a great loss of energy between these two trophic levels as there is later when carnivores prey upon herbivores. Some of the energy which is transferred between the different trophic levels is returned to the system by the decomposition of plants and animals, and in the absence of human interference or major environmental change, the ecosystem is in equilibrium, powered only by insolation.

The idea of a loss of energy at each trophic level is clearly applicable to farming. Four basic food chains encompass nearly all types of agriculture (Figure 11). In type A tillage crops are consumed directly by man, so that the food chain is short, and output per hectare of human food is high. In contrast in type C

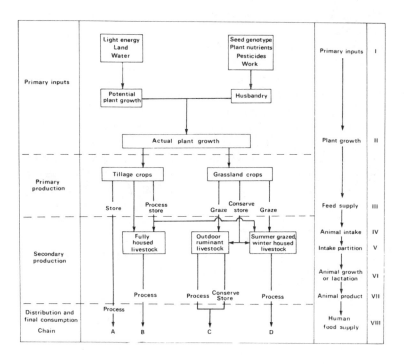

Figure 11   *The four food chains*
     Source: A. N. Duckham and G. B. Masefield, *Farming Systems of the World* (London, 1970), p. 9.

livestock graze upon grassland or natural vegetation, and man eats livestock products such as milk or meat. There is a considerable loss of energy because of the addition of an extra trophic level, and so type C only produces between 3 and 50 per cent of the amount of human food per hectare that type A produces.[4] This range reflects the varying intensities with which grazing is pursued for it encompasses both nomadic pastoralism and intensive dairying in New Zealand.

The difference between the productivity of livestock and arable farming is further illustrated by the food output per hectare obtained upon a British university farm (Table 4). Cereals produce six times the amount of edible calories as milk, the most efficient animal product, and twenty-two times as much as fat lambs. Thus it takes a far greater area to produce a given calorific value from animal products than from most food crops. This explains the marked difference in the prices of animal products and crops that has persisted throughout European history. It should be also noted that food crops also produce more protein per hectare than livestock products.

These ratios explain the great preponderance of cereals in modern world agriculture, and the dominance of cereals for direct consumption in Europe before the industrial revolution. In traditional societies low incomes mean that the demand for livestock products is limited, while the higher the population density the less the land that can be devoted to livestock. However these remarks need some qualification, for because of the large areas needed for livestock products we would expect to find far fewer livestock in Western Europe in the past than there actually were. First, livestock were not kept only for milk and meat; sheep also produced wool, while cattle gave hides, fat and tallow. Second, cattle were kept primarily as draught animals; although the horse was used as a draught animal from the twelfth century, oxen remained the dominant means of draught on many European farms until the nineteenth century. Third, livestock could use land which could not be used for crops. Cattle and sheep – ruminants – can digest the fibrous parts of plants which are inedible by man and convert them into meat or milk. In medieval and early modern agriculture few farmers could spare land to grow fodder crops specifically for livestock. Livestock however could maintain themselves upon the poor grassland of the commons, often unusable for crops, on the straw stubble left after harvest, and in the case of sheep on meagre upland vegetation. Pigs and poultry, which are not ruminants, depend for their food upon plants that can also be directly consumed by man; they thus relied largely upon the residues of foods consumed by man, and so were rarely numerous before the nineteenth century. Lastly, livestock were an important way of restoring plant nutrients to the soil before the adoption of artificial fertilizers; indeed as late as the 1850s and 1860s cattle were kept by many farmers in eastern England for the sake of their dung as much as for their meat or milk.

As long as income per caput remained low in Europe, and the cost of moving agricultural produce was high, livestock were rarely kept specifically for milk or meat, and grains occupied the greater part of the cropped land. There were exceptions of course. Where grass yielded well and cereals poorly, there were specialized livestock areas; this was noticeable in the wetter parts of western Britain. Similarly many upland areas had soils too poor for cereals, but sheep could wrest a livelihood from the poor grasses on limestone and gritstone soils. The economic transformations of the nineteenth century altered this. The growth of income per caput increased the demand for meat and milk, while the falling cost of importing grains and other crops from overseas made it possible to feed livestock with grain, a quite fundamental break with the past.

Table 4   *Output of edible calories and protein on Reading University farm*

	*Dietary kilocalories/ha**	*Dietary protein (kg/ha)**
Main crop potatoes	7324	458
Sugar beet	6733	–
Cereals	5183	260
Early potatoes	4122	257
Milk	884	101
Eggs	448	76
Semi-intensive beef	312	36
Poultry	290	85
Livestock for fat lamb	227	20

* Includes allowance for imported feedstuffs.

Source: A. N. Duckham and D. H. Lloyd. 'Production of dietary energy and protein on University of Reading farm'. *Farm Economist*, vol. 11 (1966), pp. 95–7.

The calorific output of the major crops

Not only are there major differences in the food output capacity of livestock and food crops, but there are also quite wide differences in the calorific output per hectare between the major food crops. Crops differ in the yield of dry matter they produce; but they also differ in the amount of the harvested material which can be eaten and vary greatly in their calorific value. These facts are of some importance in the interpretation of agrarian history.

The difference between the potato and cereals is perhaps of most interest in terms of European agrarian history. Potato yields are greater than those of wheat, and whereas nearly all the harvested potato can be consumed, wheat has to be milled. Currently potatoes produce 40 per cent more calories per hectare than wheat, and it is likely that this ratio was greater in the past. In the late eighteenth and early nineteenth centuries, contemporaries believed that in Ireland the calorific value per hectare of the potato was two to three times that of wheat. Although the potato was first introduced into Western Europe from the Americas in the late sixteenth century, it was not widely adopted – except in Ireland – until the early nineteenth century. This was partly because of the difficulty of adapting the crop to a very different environment, but it was also disliked as a food crop and was

Table 5 *Calorific output of African food crops, 1969–71*

	Yield (kg/ha)	Calories/100 g	Index of calorific yield
Millets & sorghum	630	345	100
Maize	1179	360	195
Rice	1830	359	302
Yams	6000	90	248
Sweet potatoes	5954	97	265
Plantains	8000*	75	283
Manioc	8000	109	337

* 1948–52.

Sources: B. F. Johnston, *The Staple Food Economies of Western Tropical Africa* (Stanford, 1968), p. 93; FAO *Production Yearbook 1957*, vol. 11 (Rome, 1958), pp. 33–48, 75–7, 103; *Production Yearbook 1979*, vol. 33 (Rome, 1980), pp. 96–107, 114, 116.

Table 6   *Calorific output of Asian food crops. 1969–71*

	Yield (kg/ha)	Calories/100 g	Index of calorific yield
Millet & sorghum	640	345	100
Maize	1817	360	296
Rice	2338	359	374
Sweet potatoes	8247	95	354
Bananas*	12,000	75	407
Manioc	9287	109	456

* 1948–52.

Sources: as Table 5.

regarded as fit only for feeding livestock. Rising prices for grain – particularly during the Napoleonic Wars – seem to have spurred its more widespread adoption, and by the 1840s it occupied a considerable proportion of the arable area not only in Ireland, but also in Norway, Belgium, the Netherlands, the Rhine valley and parts of France.[5]

Similar changes took place in Africa in the nineteenth and early twentieth centuries. The indigenous cereals – millet and sorghum – had a low calorific yield (Table 5) and were supplanted by the imported manioc – a root crop – and maize. The high yields for rice in Africa are a very recent feature, and rice was not formerly important except in the south west of West Africa. In Asia, by contrast (Table 6), although the root crops and bananas give higher yields than maize, wheat or millet, rice gives yields almost comparable with the roots and did so before the introduction of high yielding varieties. Rice has thus only rarely been replaced by root crops; however on sloping land where wet rice could not be easily cultivated imported American root crops, such as sweet potatoes and manioc, spread rapidly in China in the eighteenth century and in Java in the late nineteenth century, for they gave a much higher calorific yield than the cereals which could be grown on sloping land, including dry rice.[6]

The nutrient cycle

Modern research by ecologists has shown how the chemical elements required for life cycle between plants, animals, litter and the soil, so that in a mature ecosystem the nutrient supply is in equilibrium. There is little loss except by leaching. However once man begins to farm the land, and harvests crops or the products of animals, plant nutrients are removed from the cycle and some means of replacement is necessary. If this is not done then crop yields may decline and soil erosion may occur. Thus farmers have tried a wide variety of means of replacing plant nutrients.

One means of maintaining soil fertility is still widely practised in shifting agriculture in the tropics. The natural vegetation is cleared by use of the axe, cutlass and fire, and a variety of crops sown in the patch cleared; after two or three years the land is abandoned, and the natural vegetation allowed to regenerate. Various studies suggest that after periods varying between fifteen and twenty-five years the secondary forest cover will restore fertility and the land can be cropped again. There have been few studies of the time taken for fertility to be restored by 'natural fallow' in temperate Europe, although nitrogen does accumulate rapidly under grassland or woodland.[7] Some have argued that shifting agriculture was formerly the prevailing method of farming in Europe, and it was only replaced when population growth prevented a long enough natural fallow. By the twelfth or thirteenth centuries it was rarely found, although it persisted in more sparsely populated areas in Scandinavia and Russia until the nineteenth century.[8]

By the twelfth century the two field or three field system was the most common method of farming in most of Western Europe. In the latter, two years in crops were followed by a year in fallow, in which no crops were sown; the land was ploughed during the fallow, but not always weeded. The values of the fallow were several. First it prevented the build-up of soil-borne plant disease. Second it gave an opportunity to rid the land of weeds. Third it allowed an opportunity for nitrogen to accumulate in the soil. The biological efficiency of the open field system has been a matter of much debate. At the beginning of this century many historians believed that continuous cropping of wheat in medieval times had depleted the soil of plant nutrients, and by the fourteenth century crop yields were decling. The fall in population due to the Black Death and subsequent outbreaks of bubonic plague led to a spread of grassland farming which allowed, under a continuous pasture cover, the restoration of soil fertility. Such an interpretation has recently been restated in an ecosystems context.[9]

Certainly the open field system had upper limits to its productivity.

Land in the arable fields was always in crops or bare fallow, and never put down to grass; legumes other than the pulses – peas, beans and vetch – were not grown, and the latter were a negligible proportion of the arable before the fourteenth century. Indeed as late as 1800 they occupied no more than 4 per cent of the land in crops in Western Europe.[10] Manure is thought to have been a major source of plant nutrients; farmyard manure contains nitrogen, potassium and phosphorus, the three essential plant nutrients. But livestock were kept mainly upon the common grazing land surrounding the arable; their dung was only directly applied to cropland when they were fed upon the stubble after the grain harvest, or 'folded' – kept in temporary enclosures on the fallow. The method of making farmyard manure by feeding cattle in stalls with fodder brought to them was very unusual in medieval times, although not entirely unknown. Thus some would argue that livestock dung was a very minor source of plant nutrients in medieval farming.

Two attempts have been made to estimate the flow of plant nutrients through the medieval farming system, both confining their attention to the flow of nitrogen. R. S. Loomis[11] has argued that with the low cereal yields of the Middle Ages – about 1000 kg/ha – only 20 kg/ha of nitrogen was removed from the system at harvest. But this was replaced: rain, dust, and bird droppings would provide 8 to 12 kg/ha, the seed for the crop 4 kg/ha, assuming a seed/yield ratio of 1:4; manure from draught oxen would provide 5 kg/ha, weeds on the fallow and in the growing crop 2 to 10 kg/ha. A further input came from the biological fixation of nitrogen in the soil. There are two ways in which micro-organisms in the soil fix nitrogen. Most important are the *Rhizobium* bacteria found in association with the roots of legumes; but there are also free-living organisms, independent of legumes. The latter according to Loomis could fix 2 to 5 kg/ha. Assuming that 10 per cent of the cropland was under legumes, a further 20 kg/ha of nitrogen would be fixed each year. Even if the latter is discounted, the return of nitrogen to the soil was sufficient to maintain yields, and there is no reason to suppose that crop yields would have declined.

A very different calculation of the nitrogen cycle in traditional agriculture, in the three field system in the eighteenth century in Western Europe, has been made by G. P. H. Chorley,[12] who like Loomis concludes manure returned relatively little nitrogen to the soil in the absence of any large area under legumes. He believes that the balance was maintained by biological fixation of nitrogen, not by bacteria associated with legumes but with free-living organisms. Whereas Loomis put this figure at 2 to 5 kg/ha Chorley notes that recent agronomic work suggests that blue-green algae, whose nitrogen fixing in the tropics has long been recognized, also make an important contribution in temperate areas.

It thus seems that the nutrient cycle in the medieval system would support constant cereal yields, albeit at a low level. This is confirmed by what little information there is upon crop yields in the thirteenth, fourteenth and fifteenth centuries, which show no sign of decline. Further where there have been long run studies of continuous cropping upon experimental farms – at Rothamstead and in North America – it can be shown that in the absence of manure or fertilizers yields initially decline and then stabilize. The introduction of a fallow increases yields.[13]

Thus the medieval farming system was capable of maintaining yields at a low level, but not of increasing them. Subsequent increases in crop yields depended upon reducing the loss of plant nutrients in the cycle, or of introducing nutrients from outside the system. A number of methods were adopted. One was to alternate crops and grassland in rotation. Under this method of husbandry nitrogen accumulated in the soil while the land was under grass, and grazing livestock's dung recycled a further part of the nitrogen so that following crops had increased yields. Convertible husbandry seems to have been widely practised in midland England from the late sixteenth century.[14]

More important was the growth of legumes, whose associated bacteria fix nitrogen at a much higher rate than grassland, or than the free-living bacteria. Leguminous agricultural plants are few, and in Western Europe can be divided into two – the pulses, peas, beans and vetch – which are grown for human food, but may also be fed to livestock; and the plants sown with grass to provide swards for grazing livestock – clover, trefoil, lucerne and sainfoin. Peas, beans and vetch are harvested for their seed, so that only the nitrogen in their roots remains in the soil; the fodder legumes however contribute not only the nitrogen from plant residues, but that which circulates through grazing livestock and returns as dung; there is clearly a difference in the effect upon crop yields (Table 7).

The pulses seem to have occupied a very small proportion of the

Table 7   *Nitrogen fixation and cereal yields*

	N fixed by legumes (kg/ha)	Yield of cereal (kg/ha)
Lucerne	504	2802
Clover	291	2406
Sweet clover	302	2370
Soybeans	179	1480
Field beans	78	1300
Cereals every year	–	1080

Source: E. J. Russell, *Soil Conditions and Plant Growth* (London, 1961), p. 34.

arable land in the Middle Ages, although there was some increase in the fourteenth century; as late as 1800 they are thought to have occupied no more than 4 per cent of the arable land of Western Europe. It was therefore the spread of clover, lucerne, sainfoin and other legumes sown in swards that was of critical importance, for their cultivation in rotation with cereals increased the fodder supply for livestock and the nitrogen content of farmyard manure. Turnips and sugar-beet residues in contrast provided energy for livestock, but did not improve the nutrient content of manure. According to G. P. H. Chorley the fodder legumes were not important in Western Europe before the 1760s, but by 1880 together with the pulses they occupied about 12 per cent of the land in crops. This increased the rate of fixation of nitrogen in the soil and the supply of nitrogen in farmyard manure, and was largely responsible for the increase in cereal yields which took place between the 1770s and 1880s. It is of significance that recent work on the spread of clover in England suggests it predates the West European adoption of this innovation. It would seem that clover spread rapidly from the 1660s and was well established by the later eighteenth century. Yield increases may then predate those in Europe.[15]

The medieval system and the growth of legumes were closed systems. But in the nineteenth century farmers began to import fodder crops grown elsewhere to feed their animals; by the 1960s one half of Europe's cattle feeds were imported from outside the continent.[16] Similarly the supply of plant nutrients was increased by the import of fertilizers from elsewhere. In the nineteenth century guano was brought from Peru; but in the twentieth century the Haber process has allowed the fixation of nitrogen and the mass production of nitrogen fertilizers, which are responsible for a considerable proportion of the yield increases obtained since the 1930s.

From closed to open agro-ecosystems

Until the nineteenth century both the maintenance and the increase of crop yields required manipulation by the farmer, but using resources on the farm. But from the nineteenth century plant nutrients were imported from off the farm in the form of animal feeds and artificial fertilizers; thus in effect the agro-ecosystem of one region was supplemented by nutrients derived from another part of the world. A similar change has taken place in the source of work energy on the farm. In traditional agriculture this was provided by human and animal muscle, both powered by food and so ultimately from photosynthesis. But in this century not only has the use of imported feeding stuffs and artificial fertilizers increased

Table 8 *Food energy output per man-hour of farm labour*

Agricultural system	Output (megajoules/man-hour)
A *Pre-industrial crops*	
Subsistence rice, tropics	11–19
Subsistence maize, millet, sweet potato, tropics	25–30
Peasant farmers, China	40
!Kung Bushmen, hunter gatherers	4.5
B *Semi-industrial crops*	
Rice, tropics	40
Maize, tropics	23–48
C *Full industrial crops*	
Rice, USA	2800
Cereals, UK	3040
Maize, USA	3800
D *Full industrial crops plus animal*	
Sheep, cattle, pig and poultry, dairy UK	50–170
Cereal farms, UK (small animal output)	800
E *UK food system:*	30–35

Source: G. Leach, *Energy and Food Production* (Guildford, 1976), p. 9.

in the developed countries, but human labour has been supplemented by petroleum and electricity. It is possible to measure the energy input from petroleum, fertilizers, feeding stuffs and other energy uses on the farm in megajoules, and also to measure food output in the same units.

It can be seen that farmers who use few of these inputs (Table 8) have a low energy output per man-hour. Where crop production is carried on with the full range of industrial outputs, output per man-hour is thirty to forty times that in pre-industrial farming communities. However where crops *and* livestock are produced using industrial inputs, the superiority is less pronounced (Table 8). In modern industrial societies much of the food produced on the farm is then processed and delivered to the retail outlets. If the energy consumed in this process is counted, then the output per man-hour of the United Kingdom food system – farm, factory, wholesaler and shop – is much the same as that of a subsistence society (Table 8).

Table 9   *Energy ratios*

Subsistence, cassava crop	60–65
Chinese peasants	41
Tropical crops, subsistence	13–38
Tropical crops, some fertilizer, machinery	5–10
Sugar-beet, UK	4.2
Wheat, UK	3.35
Maize, USA	2.6
Barley, UK	2.4
Maize, UK	2.3
Potatoes, UK	1.6
Rice, USA	1.3
Peas, UK	0.95
Milk, UK	0.37
All agriculture, UK 1968	0.34
All food supply, UK 1968	0.2

Source: G. Leach, *Energy and Food Production* (Guildford, 1976), p. 8.

If energy inputs are related to energy outputs then a striking conclusion is reached. Pre-industrial societies, with no industrial inputs, have energy ratios in the range of 1:13 to 1:65 (Table 9); in contrast the major crops grown in the United Kingdom and the USA have energy ratios varying between 1:0.95 and 1:4.2; in 1968 British agriculture required 1 unit of energy input to produce 0.34 units of energy output, and the total food supply system needed 5 units of input to produce, process and deliver 1 unit of food energy.

Conclusions

The biological aspects of agriculture have received comparatively little attention from historians and geographers, and the idea of an agro-ecosystem is only just beginning to be used to illuminate world patterns of agricultural geography. Historical studies have tended to emphasize the destructive role of man in the agro-ecosystem. Thus for example archaeologists have shown how over-irrigation led to the increasing salinity of soils in Mesopotamia, the reduction of crop yields, and the subsequent demise of the civilization of Sumer.[17] Numerous studies have shown how shifting agriculture can degrade the vegetation so that forest never recolonizes the fallow, and the grassland that establishes itself cannot maintain soil fertility. But this chapter has emphasized how the ecosystem approach can throw light upon beneficial changes in agricultural history.

6 Climatic change and farming history

At the beginning of this century many writers believed that changes in climate had exercised a great influence upon human affairs. Perhaps the most notable of such writers was the American geographer Ellsworth Huntington; but in the reaction against geographical determinism, such ideas lost currency, and it is only in the last decade or so that there has been a revival of interest in climatic change, and its influence upon the history of land use and rural settlement.

The measurement of past climates

There are various ways of tracing climatic change over the past two or three millennia. By far the most important are instrumental records; unfortunately those kept by individuals or organizations do not go back very far. The oldest English temperature records date from 1684, while records have been kept at Utrecht since 1715. But there are difficulties of interpreting these data as they were not taken under standardized conditions. Phenological records – records of dates at which plants leaf or ripen – have been used to indicate climatic trends. Thus Robert Marsham and his descendants kept annual records in Norfolk of the date of the leafing of six trees from 1745 to 1886, allowing the earliness and lateness of spring to be traced. Better known is the work of Emmanuel Le Roy Ladurie, who collected the dates at which French vine harvests took place back to the sixteenth century. He argued that the earlier the harvest, the warmer the summer. His views however have been much criticized.[1] Records of the freezing and thawing of ice have been used to construct temperature fluctuations. Thus the length of time sea-ice existed around Iceland has been used to reconstruct temperature trends in that country from AD 900, while records of the number of days a Dutch canal was frozen gives a measure of winter severity from 1658 to 1839. Observations on the freezing of the Thames and Lake Malaren in Sweden have been used to establish periods with very cold winters. More recently ice cores from Greenland have been analysed and differences between the heavy oxygen isotope and that in standard ocean water have been used to infer temperature changes. In some parts of the world differences in the size of tree rings – the annual growth – have been used to measure relative aridity in dry climates and the relative coldness of a year in humid regions. Over much longer periods pollen analysis can indicate what plants were growing, and from this inferences about the climate can be made. In the mountain regions of Europe the advance and retreat of glaciers gives some indication of long run temperature changes. Lastly some authorities believe that references to climatic extremes in medieval chronicles may be used to construct measures of winter severity and summer wetness.[2]

Trends since the Pleistocene

The end of the Pleistocene is conventionally put at 8000 BC (Figure 12). This was followed by a period of rising summer and winter temperatures, reaching a climatic optimum 5000–3000 BC when the average summer temperature in England was about 18°C compared with the present 16°C. Northwards sea-ice retreated, and southwards the Sahara was wetter than at present. The climate of north western Europe then deteriorated, culminating in an Iron Age cold period 900–300 BC, which was also wetter than at present. Since then fluctuations of temperature have been smaller but climatologists claim temperatures recovered to give a secondary optimum AD 1000–1200, when summer temperatures were higher than at present. This was followed by the little Ice Age, sometimes dated from AD 1430 to 1850, which was colder than the first half of this century. In the northern hemisphere above 50°N latitude mean annual temperatures were some 1–3 per cent below the present. In the late nineteenth century, there was a rise in annual temperatures and a marked retreat of glaciers in Europe; the world mean annual temperature rose 0.6°C between 1900 and 1940, but there has been a decline in the northern hemisphere since 1940.[3]

The most striking feature of this pattern is the smallness of fluctuations in the last 2000 years compared with the preceding 8000 years. *A priori* one might expect there to be greater agrarian

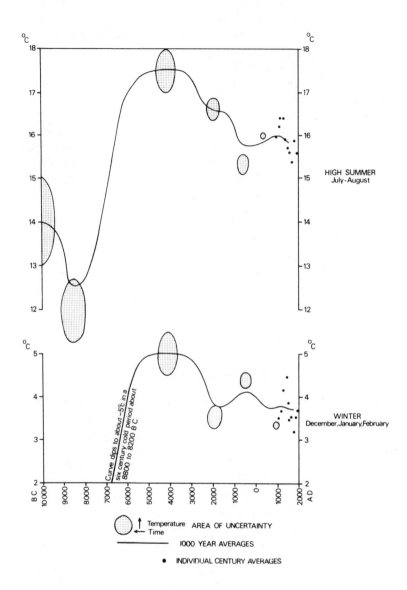

Figure 12   *Trends in winter and summer mean temperatures since the Pleistocene*
Source: H. H. Lamb, *Climate, Present, Past and Future*; vol. 2, *Climatic History and the Future* (London, 1977), p. 39.

responses to climatic change in the latter period than in the more recent millennia. Indeed some archaeologists believe that the domestication of crops and animals, the most dramatic of all agrarian changes, was caused by climatic change at the end of the Pleistocene. However the changes over the last 1000 years have been comparatively small; H. H. Lamb estimates there has been a fluctuation of 2°C in temperature and 10 per cent in rainfall in England. Many historians believe this to have been too small to have influenced long run trends in agriculture. Nor are they happy about the means by which long term trends in climate have been constructed. Descriptions of climatic events by the medieval chroniclers were almost certainly exaggerated, and their value was limited by the shortness of their experience of abnormal events. Historical records are rarely available for one place for a long run, and evidence has to be patched together from records of places often quite far apart. Many measures of climatic change refer to only part of the year, and not perhaps to a season with agricultural significance. Thus thawing or freezing in winter may have no effect upon crops sown in autumn or spring. Further, even when changes in agriculture can be correlated with climatic change, they may be often as plausibly explained by economic or demographic events.[4]

How does climate affect agriculture?

The relationship between climate and agriculture is, as was noted earlier (see pp. 48–9), difficult enough to establish when accurate records of climate and crop yields are available. Modern research has established only a few such correlations in England. Heavy autumn rainfall makes it difficult to work the land; an inadequate seed-bed will give reduced yields, and in a very wet autumn the land will have to be left fallow. As wheat is generally sown in autumn in Western Europe this could have disastrous consequences and this was possibly one cause of the great famine of 1315–17. Very low temperatures in December–February have very little influence upon crops, but a cold winter and a cold spring would reduce the fodder supply and lead to heavy losses of sheep and lambs in upland areas. A late and cold spring does have adverse effects, for it delays the sowing of spring crops, reduces the period in which the crop is receiving insolation, and thus reduces yields. A cold spring also reduces grass growth and can affect milk yields.

Cereal yields appear to be inversely related to rainfall in June, July and August, and thus wet summers adversely affect yields, while a wet harvest delays ripening and knocks down the crop, making it costly to gather in. A wet warm summer encourages the spread of parasites in livestock and promotes the spread of crop diseases such as the potato blight.[5]

Given that climatic change *could* influence crop yields and also the type of crop that can be grown, it is still difficult to establish correlations between climatic change and agricultural change.

In the first place a distinction between long run and short run fluctuations should be noted. It is not doubted that harvests in pre-industrial Europe were profoundly affected by the weather. J. Z. Titow has demonstrated from the account rolls of the manors of Winchester that poor harvests between 1209 and 1350 were related to two distinctive weather sequences: a wet autumn followed by a wet winter and summer; and a wet autumn followed by an average winter and a dry summer. W. G. Hoskins has argued that wheat prices are inversely related to the size of the harvest. In a study of wheat prices from 1480 to 1759 he claimed there was no long run trend in the frequency of bad harvests, but that there was a tendency for bad harvests to occur in spells of three or four consecutive years. This he saw to be the result of farmers consuming part of their seed-corn after a bad harvest and thus being unable to sow the same acreage the following year.[6]

In the second place there are no reliable records of crop yields in England before 1885 or of land use before 1866. Instrumental records of temperature on a monthly basis are only available since the 1680s and of rainfall since 1715. It is thus difficult to establish a relationship between climatic trends and agricultural change over a long period. Further, in the modern period the yields of crops have been profoundly influenced by fertilizers, pesticides and herbicides, and these have been the major cause of trends in crop yields, not weather.

In the third place it is debatable if the small long run fluctuations in the English climate have been sufficient to influence yields. Some modern instances do however suggest that they may have been. Thus for example changes in temperature have influenced the length of the growing season, that is the number of days between the last frost in spring and the first in autumn. The growing season in central England 1691–1700, during the little Ice Age, was 220 days; at Kew in 1931–40 it was 280 days. The fall in temperature 1940–70 has reduced the growing season by two weeks in central England. In the absence of any changes in farming methods these changes in the amount of energy received in the growing season would affect yields. In the United States two interesting studies of recent weather have been made. Over the last thirty years at six stations in the corn belt the growing season has been reduced by 14 per cent, an average of 27 days, which would be sufficient to reduce yields. But over the same period there has been a great increase in fertilizer use and the introduction of hybrid varieties, and corn yields have tripled. A mathematical model based on temperature, rainfall and technology has demonstrated that in the United States spring wheat belt a change of 1°C in each month of the growing season would be sufficient to cause a fall of 7.5 per cent in yield; but this small fall in yields would have a much greater effect upon profits. It would be generally agreed that the small fluctuations in temperature and rainfall that can be established over the last 1000 years would have a more pronounced effect upon marginal areas. Thus in Iceland a fall in the mean summer temperature of 0.8°C has reduced the growing season by two weeks, and the number of degree-days – a measure of total energy received – by 27 per cent, which in turn reduced hay yields by 25 per cent between the early 1950s and 1966–7. Comparatively small changes in upland areas may have a greater than expected influence upon land use. Thus in upland areas mean annual temperatures decline with altitude, limiting the range of crops that can be grown. Although the decline of temperature with height is linear, the probability of harvest failure increases much more rapidly.[7]

It is difficult to come to any firm conclusions on the possibility that long term trends in climate have influenced land use. It is possible that the comparatively small fluctuations in annual temperature and rainfall that have occurred over the last 1000 years had related changes in critical seasonal variables that influenced crop yields and crop selection. It is certain that these influences would have been greater in upland areas and more northerly positions, where agriculture was already marginal.

Climate and agriculture in England since AD 1000

It seems generally agreed that the climate in Western Europe in the high Middle Ages was warmer than before or after. Between

AD 1000 and 1200 summer temperatures were possibly 0.5°C higher than now, and 1.0°C higher than in the seventeenth century, with an absence of severe winters during much of the late twelfth and thirteenth centuries. Two effects of this have been noted. In the first place there is evidence of land in cultivation well above present limits in southern Scotland, Weardale, Nidderdale, Dartmoor and Snowdonia. As seen earlier (p. 58) it was population growth that caused the encroachment into marginal areas, but it was of course the milder weather that made it possible. The twelfth and thirteenth centuries also saw the northward extension of vineyards into Prussia, southern Norway and England. Again, this extension doubtless has an economic explanation, but the warmer summers made it possible; further, few vineyards were founded after 1310, by which time the climate was possibly deteriorating.[8]

The secondary optimum of the high Middle Ages was followed by a long period of cooler and wetter summers and colder winters, described by some as the little Ice Age; unfortunately neither its beginning nor its end seems agreed. Some believe it began after 1250, others about 1430, while some date it from the mid sixteenth century. It is however accepted that the nadir was reached in the very cold period of 1680–1700. The significance of this deterioration is not clear. If colder and wetter weather began after 1250 then it may have led to a slow decline in yields, which some historians have argued occurred in the late thirteenth century, although they have attributed it to soil exhaustion (see p. 34 and pp. 74–6). The higher frequency of wet autumns 1300–50 may have influenced yields and reduced the area which could have been sown to winter wheat. The cooler weather might have been responsible for the end of the expansion of vineyards and their later contraction, and to the abandonment of upland settlements. But all these events have equally plausible explanations in terms of price movements or population contraction. In the absence of any convincing evidence of climatic change these seem more likely explanations.

Some authorities believe that the period 1450–1550 saw a temporary arrest in the increasing coldness of winters in Western Europe. Thus for example there is no record of the Thames freezing between 1434 and 1540. This is also a period when the population began to recover from the outbreaks of bubonic plague that had occurred in the second half of the fourteenth century. In this period cherries, peaches, quinces and figs were first grown in

England, and upland areas were occupied once again. But climate allowed this expansion, it did not cause it. After 1550 it is thought wetter summers and colder winters became more frequent. There is some evidence from studies of cereal prices that harvest failures became more frequent after 1550 than before. Between 1300 and 1700 the growing season in England may have been reduced by about one month which could have reduced crop yields, although there is no evidence to support this.[9]

Temperature records begin in the late seventeenth century. They suggest this was an exceptionally cold period of English history, and literary evidence supports this. There was however a notable increase in mean annual temperatures in 1700–40 which then remained constant until the 1890s. The early eighteenth century also saw an above average number of dry years. These changes may have had some influence on English agriculture for there was a growth in total output, and a considerable export of grain in the first half of the eighteenth century.[10]

It is hard to discern any profound climatic influence upon English agriculture over the last millennium, and in the absence of reliable climatic records or agricultural statistics it is difficult to resolve these questions. It should be noted that the contemporary beliefs of farmers are not necessarily a guide to the true causes of agrarian change. In the 1880s British corn producers universally blamed the series of wet harvests for their undoubted misfortunes; few blamed the imports of grain from the New World which were so rapidly reducing their prices.[11]

Conclusions

In spite of the increasingly sophisticated methods of estimating climatic change, it is difficult to show unequivocally that the temperature and rainfall changes of the last millennia have had any profound effect upon long run changes in agriculture. This was largely because the technology even of pre-industrial periods was capable of adjusting to climatic change, which only altered slowly. It was not capable of adjusting however to the harvest failures that occurred in years of very poor weather. The decline of harvest failure on a large scale after the late seventeenth century marks the beginning of modern history.

Part Three
Industrialization and agricultural change

7 The nature of peasant societies

The most profound break in the long history of agriculture came in the nineteenth century when industrialization and urbanization transformed the rural world and revolutionized agriculture. These changes are analysed in the following chapters, but it may be useful to consider first some of the general characteristics of pre-industrial agricultural society and some of the ways which have been developed to analyse such societies.

Economic theory, peasants and farmers

In the nineteenth century certain assumptions were made which have become the basis of most modern economic theories that seek to explain the behaviour of farmers. J. H. von Thünen, in what was the first economic model (see pp. 135–9) assumed that the farmer's prime aim was to maximize his profits, that he was aware of changes in the price of his inputs and products, and responded rationally to these changes.[1] Later in the century marginal analysis combined with the older law of diminishing returns provided a means of demonstrating how farmers could maximize profits by responding to changes in the price of inputs and outputs; the rational farmer would increase his inputs until the marginal expenditure and revenue were equal. At about the same time ideas on the economies of scale in the organization of industry were applied to the size of farms, and were used to confirm a long held 1–although rarely explained 1–belief that large farms were economically more efficient than small farms.[2]

So a body of theory grew up to explain the behaviour of farmers, largely based on the experience of farmers in Western Europe and in European settlements overseas; but it was assumed that *all* farming communities worked on the same assumptions and could be analysed with the methods of modern economic analysis. There was however some opposition to these views. Marx had argued that different modes of production followed different laws of behaviour, and that slavery, feudalism, capitalism or the Asiatic mode of production would each require a form of economic analysis specific to that mode of production.[3] At the turn of this century a number of Russian agricultural economists argued that the peasants of Russia and Eastern Europe had different aims and obeyed different economic laws; in the 1920s A. V. Chayanov put forward a theory of the peasant household which remains the only coherent alternative model to that of the capitalist profit-maximizing farmer.[4] J. H. Boeke, a Dutch economist, subsequently applied some of Chayanov's ideas to the economy of the Dutch East Indies, and argued that there were two distinctive sectors in the economy, the peasant subsistence sector behaving in a quite different way to the commercial, capitalist sector.[5] Since the end of the Second World War some economists have argued that farmers in the developing world do not respond to price changes in the same way as Western farmers, and have stated that a different form of economic analysis is necessary to explain their behaviour.[6] Similar views have long been held by some anthropologists; they have suggested that economic behaviour in tribal and peasant society is not determined primarily by external forces such as price changes, but by non-economic factors such as obligations to kin groups or to tribal chiefs.[7] More recently there has been much study of 'peasants' who are seen as a group analytically different from capitalist farmers on the one hand or isolated tribal communities on the other.[8]

On the whole historians – even economic historians – have not used modern economic theory to interpret agricultural change in the past, although recently there have been some attempts.[9] But it is worth considering whether modern economic analysis is appropriate to farming prior to the industrial revolution. Is it not possible that the alternative views put forward by anthropologists and economists to explain modern peasant behaviour may be more applicable to farmers in Western Europe before the nineteenth century?

A. V. Chayanov's work provides the only model of peasant behaviour, and is briefly considered. Then some of the characteristics of modern peasant economies are outlined, and their relevance to the past is touched upon.

A. V. Chayanov and the theory of the peasant household

A. V. Chayanov is the best known of a group of agricultural economists who studied the Russian peasant economy in the late nineteenth and early twentieth centuries, using as evidence the great deal of statistical information on peasant communities available in the *zemstvo* reports which were recorded from the 1870s until the First World War. Chayanov argued that the peasant was fundamentally different from the capitalist farmer, so different that it could be said that a peasant *mode of production* existed. He went further and argued that peasant farming could be as efficient as capitalist farming.[10]

Chayanov argued in *The Theory of Non-capitalist Economic Systems*[11] that the peasant was not a profit maximizer; further, as his farm depended upon family labour, no wages were paid. Without wages it is impossible to calculate rent, interest or profit, as these are all interdependent. The peasant farm is based on self-exploitation of the peasant's labour and that of his family. The amount of labour input is a function of an equilibrium between the family's consumption needs, and the irksomeness of labour. This equilibrium point is determined by a variety of factors such as the type of farming practised, the environment, and distance from the market, but especially by family size and its composition. These views were elaborated in *Peasant Farm Organization* published in 1925.[12] The basis of the peasant household is the family, which hires no labour and attempts to provide its own consumption needs. The family derives income not only from agriculture, but from crafts; for much of the year family labour is unemployed because of the seasonal nature of agricultural production. He believed that there was a correlation between family size and farm size, and that this changed with the life-cycle of the family. As the number of children increased, so did the area worked, and when the children left home the farm fell in size. Chayanov recognized that this was perhaps peculiar to Russia, where the periodic redistribution of land by the *mir* continued to operate. He also recognized that his theory might not be applicable to peasants in Western and Central Europe, where hired labour was more common than in Russia.

As the peasant family grew, so the ratio between the number of consumers and producers changed; he expressed children as adult equivalents both as consumers and workers, and showed how this ratio changed as children grew up, became workers, and finally left home when they married. Thus the labour supply changes over time, but is fixed in any one year; the peasant seeks to acquire land to fit this given labour supply, either by renting, by purchase or by redistribution. Further, the peasant seeks to employ fully his given, uncosted labour supply, and to even out the work-load throughout the year. The two further determinants of peasant behaviour are family needs and the increasing drudgery. Labour inputs are increased until the family's needs are met; beyond that the increasing irksomeness of work outweighs the advantages of further production, *unless* the family wishes to increase its accustomed consumption level, or to improve the farm equipment.

Chayanov's theory has a number of implications. Although he does not regard the peasant household as a pure subsistence unit, it does not aim at profit maximization, but at satisfying the family's needs. Beyond that additional work is balanced against increasing drudgery. Marginal analysis is thus an inappropriate manner of analysing the peasant economy. Peasants will continue to increase labour inputs even if marginal returns are decreasing, until family needs are met. Chayanov insisted that the income of the peasant household was not solely from agriculture, for handicrafts might be practised at home, while near towns the peasant might work in industry for part of the year. He also believed that the differences in farm size within a village were not simply a result of the concentration of land holdings by the incipient capitalist farmers, but were a result of the life-cycle of the peasant family, with its acquisition and subsequent shedding of land holdings.

Chayanov himself recognized that his theory might not hold in those peasant areas where hired labour was important; nor could it be relevant in densely populated areas with a rigid tenurial system, for it would be difficult for a family to add land as the size of the family grew. His views have only recently become known in the West, and there are few instances of the application of his ideas to modern or pre-industrial peasant societies.[13]

The definition and criteria of the peasant household and the peasant economy

Many workers have attempted to define the characteristic features of the peasant household and the peasant economy. At present approximately half of mankind lives in such economies, and it is agreed by most writers that such communities break down with the spread of economic modernization and capitalist organization. It is the task of the following chapters to trace the impact of industrialization upon agriculture; it is the task of the rest of this chapter to establish the general characteristics of peasant farming before the industrial revolution.

Subsistence and the market

All writers have agreed that the peasant aims at providing as much of the family's needs as possible; only a residual of produce is sold off the farm. Nineteenth-century historians laid great stress upon the evolution of the commercialization of agriculture and the decline of the subsistence economy. In fact it is extraordinarily difficult to measure the degree of subsistence in modern peasant economies, even more so in the past. J. W. Mellor has argued that in modern peasant economies typically two thirds of output is retained upon the farm, one third sold off. At the turn of this century over a quarter of the produce of Swiss farms was consumed on the farm, in the 1930s two thirds of Bulgarian output and in Yugoslavia half the crops and 40 per cent of livestock produce.[14]

The decline of subsistence agriculture in Western Europe has not been traced except in general terms, but it seems likely that some produce was sold off the farm from a very early date, if only to obtain cash to pay rent, taxes or, in some cases, *tithe*. Thus peasants would be partially integrated into the market from an early date; but some modern writers have argued that subsistence farmers do not respond rationally to price changes, either because they produce only for their own consumption, or because production is allocated according to local needs or kinship obligations. However the bulk of empirical studies supports the idea that modern peasants are integrated into the market and do adjust their crop acreages according to relative price movements.[15] Further, T. W. Schultz has argued that traditional farmers in modern India make the best use of the resources at their disposal and allocate their resources efficiently.[16] Crop yields are low not because of the incompetence of the farmer but because of the lack of modern inputs such as fertilizers or improved crop varieties. It may be that it is worth examining the farm practices of pre-industrial farmers in Europe in this light rather than assuming, as many contemporary pamphleteers did, that they were inefficient and backward.[17]

Nonetheless not all economists are convinced by the studies that show the modern peasant to be price responsive. Most of the studies which relate acreage changes to price movements have dealt with cash crops such as cocoa, jute, cotton or oil-palm, which would not be consumed on the farm. It does not follow that the producers of food crops react in the same way. Further, most studies have dealt with aggregated data. When data at the farm rather than the regional level have been analysed it has been shown that the smaller the proportion of output marketed the less the response to price change; that small farmers are less responsive to price changes than large farmers; and that producers of food crops are less responsive to price changes than cash crop producers.[18]

The peasant household

According to Chayanov the chief feature of the peasant household is that it is both a production and a consumption unit and that the domestic and agricultural functions are inextricably interlocked. He emphasized that agriculture was not the only source of income in the peasant household; indeed in parts of Russia in the early twentieth century crafts and off-farm activities occupied as much of the household's time as agriculture, while in sixteenth-century England many cottagers and small farmers also worked in forestry or the spinning and weaving of flax, hemp or wool.[19] But the peasant household also undertook many tasks that in modern farming communities are done by specialists. Jan de Vries has shown that at the beginning of the sixteenth century Dutch farmers built their own houses and farm implements, transported their produce to market, dug peat and cut reeds. During the sixteenth century these became the duties of the landless, who carried out the crafts more efficiently than the farmer, and allowed the farmer more time on purely agricultural activities; thus both farming and rural services became more productive as a result of increased specialization.[20]

The peasant and the labour supply

Chayanov and others have argued that a characteristic feature of the peasant economy is that the labour supply consists of the peasant, his wife and his children, and that hired farm labour is rare. In some modern peasant societies this is true. In modern Poland 80 per cent of private farms use only family labour; but it is not true of other societies, which on most criteria would be described as peasant. Thus in Bavaria in the 1920s, although family labour provided a great majority of the hours worked, the larger farms hired labour, as they do in modern India. In Norway and Sweden in the early nineteenth century labourers were found on even small farms, and hired labour was important in English peasant communities in the medieval period. However as late as 1851 40 per cent of the farms in England and Wales employed no labour other than that of farmer, wife and children (see p.208).[21]

The implications of the dominance of family labour have already been noted; labour was uncosted and marginal analysis is unlikely to be an appropriate way of explaining the peasant's motives (see p. 93). Further the peasant, unlike the capitalist farmer, does not shed labour in bad times or as a result of the adoption of labour-saving machinery. The existence of a labour surplus will in fact retard the adoption of innovations that save labour. Indeed at times much of Europe must have been characterized by massive *underemployment* on peasant farms; England in 1340 and Ireland in 1840 are said to have been such cases.[22]

Inputs and the peasant farm

If the peasant produced primarily for his own consumption needs, and thus sold little off the farm, he also purchased few inputs. It has been seen that little labour was hired; a wife and children are essential to the working of the farm. But few other inputs were obtained from off the farm. Seed for the following year was retained from the harvest, so as much as a quarter of the harvest had to be kept. Power on the farm came mainly from human labour and the draught power of oxen or horses. Fertilizers were confined to the use of animal manure, although in Flanders and northern Italy urban wastes and horse dung were purchased from late medieval times. Farm implements were simple, locally made by craftsmen or even by the farmer himself.

The major inputs on the pre-industrial farm were thus labour and land; capital inputs were relatively unimportant. Land thus assumed a great importance in the peasant economy. Such savings that the peasant could make would be spent upon acquiring extra land, even if remote or of poor quality, rather than purchasing inputs which might improve productivity. Systems of inheritance and the provision of dowries were often devised with the aim of retaining land within the family. Household size and the type of family – extended or nuclear – were often determined by the relative abundance or scarcity of land in the peasant community.[23]

The peasant and the village

In most parts of the world the peasant lives in a village; in the past the peasant also lived in a village, although some recent writers believe this was only so from the end of the first millennium AD. Before that the isolated dwelling house or the hamlet were more common. The village form of settlement and the associated open arable fields and common grazing land had a number of implications. First, communal labour and co-operation in seasonal tasks was more common than in modern times, where such labour needs would be met by hired casual or seasonal labour. Second, the use of common grazing and the arable fields was governed by regulations determined by the community as a whole and administered by a small group of villagers. The individual's decisions about farming were thus constrained by the decisions of a much larger group.

The peasant was also constrained by external forces outside the village. Anthropologists have insisted that a peasant economy is distinguished from tribal communites, which are more self-contained, by partial integration into a wider regional or state culture; the peasant, some have insisted, was subordinate to these authorities who were normally to be found in towns.[24] Thus landlords, the Church and the state all extracted income and labour from the peasant; it is not clear that this has much analytical significance. After all the capitalist farmer of modern England pays rent, income tax, rates and tithe, and the collective farmer of a socialist economy is compelled to deliver a proportion of output to the state. It is true however that much of the change in peasant societies, and particularly those forces that have transformed peasant economies into modern capitalist economies, have come from outside peasant society.

The peasant, risk and leisure

Many argue that modern and historical peasant societies were reluctant to take risks; and with good reason. The traditional farmer had learnt what crops could be grown in his neighbourhood, and how and when they might best be planted and cultivated. But he was still at risk from the weather or outbreaks of plant or animal disease. In such years his crop would be greatly reduced; the poor transport of the time meant that it was both costly and slow to import food. Pre-industrial society was thus characterized by crisis years, when death rates soared above the normal level as a result not so much of starvation but of the reduced resistance to infectious disease that malnutrition brought. The modernization of agriculture and improvements in transport greatly reduced such crises; there were few in England after the 1660s, and they greatly declined in most of Western Europe after 1815. In the developing world the frequency of harvest failure has diminished since the 1940s as transport and famine administration has improved.[25]

But the very high risk of harvest failure even when traditional practices were followed ensured that peasant communities were reluctant to make any change to the unknown, whether it be growing new crops or adopting new methods; the peasant could not afford failure, for failure meant not merely poverty but often death. Not surprisingly the peasant of the past, like his modern counterpart, aimed not at profit maximization but at the minimization of risk.[26]

Other writers have seen as the key to peasant behaviour not the avoidance of risk but a high preference for leisure; the peasant prefers leisure to what Chayanov called the irksomeness of labour. They thus may not increase labour inputs in order to equate marginal inputs with marginal output, but prefer instead more leisure time. This is after all rational behaviour even if it is not the rational behaviour of the nineteenth-century economist. The theories of peasant behaviour of both Ester Boserup and A. V. Chayanov depend upon this assumption.

Conclusions

The differences between the peasant and the capitalist farmer are still a matter of debate. Enough has been said to suggest that the motives and behaviour of modern peasants are not identical with those of Western farmers; it may be that the tools of modern economic theory are not those that will explain the peasant of today. It of course does not follow that the farmers of Western

Europe before industrialization behaved like the modern peasant, who is difficult enough to define; but it does suggest that some of these ideas might profitably be applied to the interpretation of agrarian change in Western Europe before the nineteenth century. Undoubtedly it is true that industrialization and the changes associated with it transformed the traditional world and slowly eroded the peasant and peasant *mores*. To these changes we now turn.

8 Structural transformation and turning points

Modern agriculture is highly dependent upon inputs from industry. Traditional agriculture was not (see pp. 97–8). Thus the industrial revolution has had the profoundest effects upon farming and has been a major cause of agrarian change in the last two centuries.

What is industrialization?

Industrialization is a complex concept, for many economic changes were associated with the growth of manufacturing industry in the eighteenth and nineteenth centuries. First, the proportion of the population employed in agriculture declined, and the proportion engaged in manufacturing and mining increased; this change is often called structural transformation. Second, there were changes in the organization of industry, notably the emergence of factories and capitalism. Third, there was a greater use of machinery, and the application of steam power. Fourth, there was an increase in the urban population and of the proportion of the total population living in towns. Fifth, there was an increase in the application of science to industry. Sixth, the industrial revolution gave economic growth; national income per caput rose, and somewhat later living standards increased. Finally, there were great improvements in transport, so that the real cost of moving agricultural inputs and produce declined.

Industrialization influenced many aspects of agrarian life. In this and the following chapters the following topics are considered: first, the consequences of structural transformation; second, the results of changes in the size of the agricultural labour force; third, the consequences of changes in demand; fourth, the consequences of industrial inputs for agricultural productivity; and fifth, the results of falling transport costs.

Structural transformation

In pre-industrial societies three quarters or more of the population were engaged in agriculture. One of the major results of industrialization has been the great decline of this proportion. Currently only 2 per cent of the work force of Britain or the United States are employed in agriculture, and agriculture produces a very small part of the national income. Indeed, in the modern world there is a very close correlation between the proportion engaged in agriculture and national income per caput. The higher the proportion in agriculture, the lower the income per caput. Although there has been a continuous decline in the proportion of the population engaged in agriculture in European areas in the last two centuries, this was not at first accompanied by a decline in the absolute numbers engaged in farming. On the contrary the typical pattern has been for the early stages of structural transformation to be characterized by an increase in numbers, followed by stagnation, and then decline. It is the consequences of these two types of change we are concerned with here.

There are unfortunately few accurate records of occupational structure before the mid nineteenth century except for Sweden and Finland, where over 80 per cent of the population were engaged in agriculture in the early nineteenth century, before industrialization began (Figure 13). In most other countries industrialization was under way before the collection of data on occupations began, but all show a continuous decline since the earliest years for which data are available, and by 1970 all were below 20 per cent. European settlements overseas have also showed a comparable decline, and since the 1950s most parts of the developing world have had a fall in the proportion engaged in agriculture, although the proportion is still much higher than in the developed countries (Figure 14). At the beginning of this century 72 per cent of the world's labour force was engaged in agriculture; it had fallen to 46 per cent in 1978.[1]

The implications of structural transformation

There are three important consequences of structural transformation for agriculture. First, it increases the non-agricultural population. Second, it changes the distribution of population and requires improvements in transport and distribution. Third, it requires

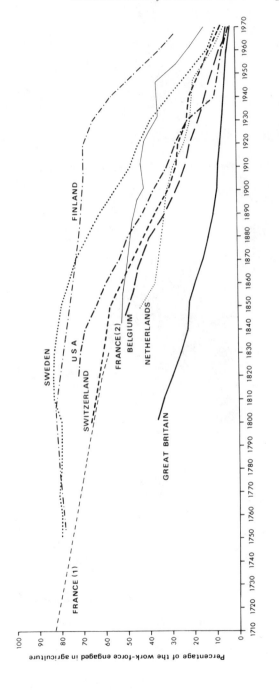

Figure 13 *Percentage of the working population engaged in agriculture in selected countries, 1710–1970*
Sources: P. Bairoch, *International Historical Statistics*; vol. 1, *The Working Population and its Structure* (New York, 1969); FAO, *Production Yearbook, 1979*, vol. 33 (Rome, 1980), pp. 66–71; J. C. Toutain, *La Population de la France de 1700 à 1959* (Paris, 1961), p. 105.

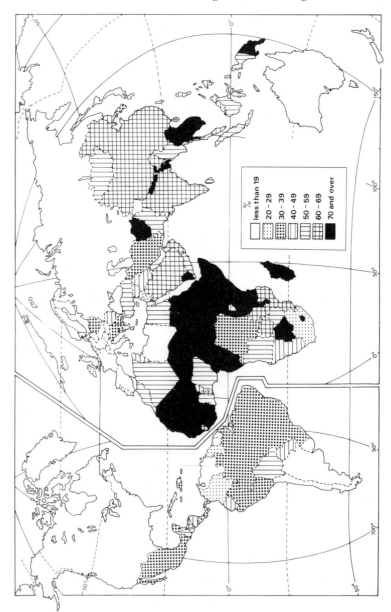

Figure 14   *Percentage of the working population engaged in agriculture, 1978*
Source: FAO, *Production Yearbook, 1979*, vol. 33 (1980), pp. 61–71.

increases in farm productivity if living standards are to be maintained. All three changes accelerate the shift from subsistence to commercialized agriculture; industrialization undermines peasant society.

When 80 per cent of the work force are employed in agriculture, there is a limited market for cash sales, as most of the population live and work on farms: each farm family has to feed only one quarter of a non-farm family (Table 10). However when the proportion in non-agriculture has risen to 50 per cent, and that in agriculture fallen to 50 per cent, each farm family has to feed one non-farm family. The market for commercial sales off the farm has tripled and triples again as the proportion in agriculture falls from 50 per cent of the total labour force to 20 per cent. This will happen even if the population is constant.

This has some interesting implications. Thus in Holland the population rather more than doubled between 1514 and 1622,

Table 10   *Structural change, market size and productivity*

(1) Percentage of population engaged in agriculture	(2) Number of non-farm population to be fed by each man in agriculture	(3) Number to be fed by each man in agriculture including himself	(4) Percentage increase in output required per man in agriculture with each 10 per cent fall in percentage in agriculture	(5) Percentage increase in number of non-agriculturalists with each 10 per cent fall in percentage engaged in agriculture
80	0.25	1.25	–	–
70	0.42	1.42	13.6	68
60	0.66	1.66	16.9	57
50	1.0	2.0	20.4	51
40	1.5	2.5	25.0	50
30	2.3	3.3	32.0	53
20	4.0	5.0	52.0	74
10	9.0	10.0	100.0	125
5	19.0	20.0	100.0	111

Source: B. F. Johnston and S. T. Nielsen. 'Agriculture and structural transformation in a developing economy'. *Economic Development and Cultural Change*. vol. 14 (1966), p. 283.

which of course increased the demand for foodstuffs. But the proportion of the population living in places of over 10,000 rose from 20 to 50 per cent of the total population, so that the commercial market for agricultural products much more than doubled. This stimulus of a growing market was no doubt a major factor in accounting for the considerable changes which took place in Dutch agriculture in the late sixteenth and early seventeenth centuries, giving rise to a highly commercialized, intensive and specialized agriculture by the 1640s. A comparable change took place in England between 1650 and 1750; although there are no occupational figures other than for 1688 and 1801 it seems probable that the proportion engaged in agriculture fell from 70–80 per cent in 1650 to 50 per cent in 1750. Although the population of England showed very little increase in this period, the demand from the non-agricultural population thus either doubled or tripled, providing an incentive to increase output that may help explain the considerable advances in farming methods that occurred in this period. In contrast France had a slow decline in the percentage employed in agriculture in the eighteenth and nineteenth centuries, which may help explain the apparently slow rate of productivity change, for a large proportion of the population remained upon the land, providing for their own needs: the absence of a rapid increase in demand for foodstuffs meant there was less incentive to improve farming methods.[2]

Structural transformation and the distribution of population

The decline in the proportion engaged in agriculture and the increase in the proportion in manufacturing industry led to an increase in the population living in towns. Before the mid eighteenth century only the Low Countries, England and northern Italy had more than a small proportion of their population living in towns of over 5000; even in 1800 only the Netherlands had more than a third of its population living in such places. By 1910 more than half the population of Germany, England, Belgium and the Netherlands were living in such towns, and in this century urbanization has spread throughout the whole of Europe.[3] This concentration of the population had consequences for farmers. It of course increased the market, and broke down subsistence agriculture; but for the process to work efficiently it needed improvements in transport, for not only did a greater proportion of foodstuffs move off the farm, but they moved longer distances to market. It was areas nearest to the great cities that first showed modern features of agriculture – commercialization, intensiveness and specialization. Thus the pioneer areas of Europe in agriculture were in northern Italy, the Low Countries and in south east England (see p. 143).

Structural transformation and productivity growth

If the process of structural transformation is to proceed smoothly, there must be an increase in agricultural productivity. As the non-agricultural population increases, so the numbers to be fed by each farm family increase (Table 10). Assuming that food consumption levels are maintained and that there is no import of food, the productivity of each farm family must increase. In the early stages of agricultural development these requirements are comparatively modest. With a fall in the proportion in agriculture from 80 to 50 per cent output per worker has to rise only 60 per cent (Table 10, column 4). In England this phase probably lasted from the early seventeenth century to the early eighteenth century. With a fall in the agricultural labour force from 50 to 20 per cent, a bigger increase in output per worker – 150 per cent – is needed. But the final phase, where the work force falls from 20 to 5 per cent, requires a tripling of output per worker. Now in England the agricultural labour force had already fallen to 24 per cent by 1831; thus without food imports the further decline to 5 per cent would have required improvements in labour productivity which were beyond the capacity of even the best farming in the nineteenth century. Food imports were thus inevitable. It is noteworthy that in nearly all other developed countries the proportion engaged in agriculture has only fallen below 20 per cent since 1930, in a period when great improvements in farming methods have made great increases in farming productivity possible.

It can be argued that the growth of the non-agricultural population is dependent upon increases in agricultural productivity, and this can be expressed in a formula.[4] The higher the rate of population growth, the more slowly the agricultural sector will decline:

$$Sg = (fi/f)(Nf/Ng)$$

where Sg is the percentage rate of increase in the proportion of the population in the non-agricultural sector, fi/f is the annual percentage increase in productivity in the agricultural sector, Nf is the proportion of the population in the agricultural sector, and *Ng* is the proportion of the population in the non-agricultural sector.

In Western Europe in the nineteenth century population growth was slow compared with the developed world today. Even so the transition to the highly industrialized societies of the present was not achieved without substantial imports of both food and animal feeding stuffs. Western Europe's industrialization did not depend solely upon its own resources.

Changes in the size of the agricultural labour force

Although industrialization leads to a continuous decline in the proportion of the population occupied in agriculture, because the non-agricultural population increases more rapidly than the agricultural population, it does not follow that the *absolute* numbers employed in agriculture decline. Indeed because industrialization has always been accompanied by population growth, the numbers in farming have typically increased, then stagnated, and only late in the process of economic growth declined rapidly. Agriculture employs all those born into it, as there is no unemployment on family farms. Employment in agriculture is thus a residual; it absorbs all natural increase minus those who migrate to the towns. This migration is a function of the rate at which new employment opportunities are created in the towns minus urban natural increase. B. F. Johnston has expressed these relationships in a formula:[5]

$$P^{1}{}_{A} = (P^{1}{}_{T} - P^{1}{}_{N}) (P_{T}/P_{A}) + P^{1}{}_{N}$$

where $P^{1}{}_{A}$ is the rate of change in the agricultural population, $P^{1}{}_{T}$ is the rate of change in the total population, $P^{1}{}_{N}$ is the rate of change in the non-agricultural population, and P_{T}/P_{A} is the weighting coefficient equal to the total population divided by the farm population.

The time when the absolute numbers in the agricultural labour force begin to decline is referred to as the turning point. The higher the percentage employed in agriculture, the more difficult it is to obtain a reduction in the numbers in agriculture and the higher the rate of population increase the longer delayed will be the turning point. D. W. Jorgenson has argued that the turning point in Europe came when the proportion employed in agriculture had fallen to about 50 per cent.[6] For Western Europe this is approximately true, the turning point coming when the proportion was between 50 and 35 per cent (Table 11) except in the Netherlands and Britain; it was not true in Eastern Europe where absolute decline began when the proportion was still high, or in the European areas overseas, when it came when the proportion was low.

As noted earlier few countries have collected figures on the agricultural labour force before the mid nineteenth century, but estimates do exist. The agricultural labour force of Sweden doubled between 1750 and 1850 (Figure 15) and there seems little doubt that it increased throughout Western Europe; rural natural increase in this period considerably exceeded migration to the towns. Even in England, where there was undoubtedly considerable rural–

Table 11   *The decline in the agricultural labour force (both sexes)*

	Date*	Percentage in agriculture		Date*	Percentage in agriculture
USSR	1926	86.1	Switzerland	1880	42.4
Rumania	1930	78.7	France	1921	41.5
Yugoslavia	1948	77.8	Sweden	1920	40.2
Bulgaria	1946	75.3	Czechoslovakia	1921	40.3
South Africa	1921	69.5	Austria	1939	39.0
Finland	1940	57.4	Germany	1907	36.8
Poland	1950	57.2	Denmark	1930	35.6
Hungary	1949	52.9	Norway	1931	35.3
Japan	1947	52.6	USA	1910	31.6
Spain	1950	48.8	Canada	1941	27.2
Portugal	1950	48.4	New Zealand	1936	27.2
Italy	1936	48.2	Argentina	1947	25.2
Luxembourg	1967	44.5	Great Britain	1851	21.9
Belgium	1866	44.4	Australia	1933	20.5
			Netherlands	1947	19.3

*Date at which absolute numbers in agriculture began to decline.

Source: D. Grigg, 'Agricultural populations and economic development', *Tijdschrift voor Economische en Sociale Geografie*, vol. 65 (1974), p. 417.

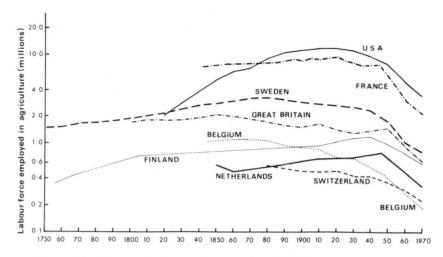

Figure 15 *Changes in the numbers employed in agriculture in selected countries, 1750–1970*
Sources: P. Bairoch, *International Historical Statistics*; vol. 1, *The Working Population and its Structure* (New York, 1969); FAO, *Production Yearbook, 1979*, vol. 33 (Rome, 1980).

urban migration in this period, the rural population doubled.[7] The turning points for Britain and Belgium came in the 1850s and 1860s respectively, and Switzerland began a slow decline in the 1880s. The first half of the twentieth century saw a slow decline everywhere in Western Europe, but this became rapid after 1945, with the labour force declining at 2–3 per cent per annum.

A quite different situation prevailed in the European settled areas overseas; the United States, Canada, Argentina, Australia and New Zealand were still sparsely populated in the mid nineteenth century, and received large numbers of immigrants from Europe later in the century. The area under cultivation continued to increase until the 1930s. The agricultural labour force in the United States reached its turning point in 1916. It was later in the other countries; indeed in Argentina it did not come until 1947.

Turning points and the agrarian consequences

During the process of economic development the proportion in agriculture continuously declines, but the absolute numbers typically increase, stagnate and then decline. It has been argued that there

Table 12 *Turning points and agrarian consequences*

Stage 1: Agricultural population increases absolutely and decreases relatively	Stage 2: Stagnates absolutely and decreases relatively	Stage 3: Decreases absolutely and relatively
1 Market for agricultural produce increases slowly	1 Market for non-subsistence agricultural produce increases	1 –
2 Mainly a subsistence economy	2 Commercial farming increases at the expense of subsistence farming	2 –
3 Peasants aim to maximize output per hectare	3 –	3 Farmers aim to maximize output per head
4 Subdivision of farms	4 –	4 Larger farmers dominate farm structure
5 Labour surplus and under-employment: landless increase as a proportion	5 Landless decline in numbers; increase in proportion of farmers	5 Acute shortage of labour: landless go to towns as do smaller farmers
6 Lack of industrial growth means few jobs outside agriculture	6 Industrial growth provides jobs outside agriculture	6 Continued labour shortage: able-bodied males and young who leave
7 Agricultural progress by means of increasing labour inputs	7 –	7 Machinery substituted for labour
8 Competition for land leads to land reform measures	8 Output fails to keep up with demand; government policies to stimulate production	8 Gap between farm and non-farm incomes widens: governments attempt to restore parity

Source: R. Bićanić, *Turning Points in Economic Development* (The Hague, 1972), pp. 155–88.

are characteristic agricultural changes in each of these three periods.[8]

Stage 1 (Table 12) requires little comment for it shows all the symptoms of agricultural overpopulation which have been discussed earlier (pp. 27–32). Acute competition for land may compel governments to pass legislation to redistribute land, as happened in Russia in the early twentieth century, in Eastern Europe between the two world wars and in many developing countries since 1950.

Stage 2 is characterized by a rapid growth of the market for agricultural produce, and commercial farming increases at the expense of the subsistence sector. The landless now have opportunities for employment in industry and migrate. Farmers thus become an increasing proportion of the total agricultural population. However agricultural output fails to keep up with demand and governments promote policies to stimulate increased production. Some of these features can be seen in the structural changes in West European agriculture in the second half of the nineteenth century. In France and Britain emigration from the land to the towns was primarily of labourers, not farmers; the number of farmers did not greatly change. Nor is there any doubt that this period saw an extension of the commercialization of agriculture and the decline of the subsistence sector (see pp. 103–6).

A third and final stage occurs when there are absolute and relative decreases, the former more rapidly than hitherto. There is now a marked exodus to the towns, not only of labourers but of small farmers who can earn more in the towns than on the land. Migration is selective, with able-bodied males predominating, so that the agricultural labour force has an increase in its female component and the average age of the farming community rises. The migration of small farmers leads to an increase in the size of farms, and the shortage of labour prompts the adoption of machinery. Farmers now aim to maximize output per head rather than output per hectare. During this period the gap between farm incomes and non-farm incomes becomes a political issue and governments attempt to restore parity.

The latter features are to be found in most West European countries since 1945; rapid industrial growth between 1945 and the late 1960s prompted a remarkable exodus not only of labourers but also, for the first time, of farmers. This has led to some increase in the size of farms and has necessitated the adoption of labour-saving machinery. However a large number of Western

Europe's farms are still too small to provide an income comparable with that obtainable in industry, and individual countries as well as the European Economic Community have pursued structural policies which encourage amalgamation and the decline of the small farm, but pricing policies that have favoured the survival of inefficient producers.

The labour force and technological change

When the labour force is increasing and wages in farming are low, there is little incentive for farmers to introduce labour-saving machinery. The adoption of the reaper in England and Europe illustrates this well. The labour force in England increased steadily from 1751 to 1851 and particularly after 1815, even although there was migration to the towns. The new farming methods required extra labour, which was provided by natural increase. After the end of the Napoleonic Wars there were signs of both unemployment and underemployment which lasted until after 1851. This was a period when eastern and southern England had surplus labour. At this time the cereal harvest was still got in with the sickle or scythe, and was threshed with the flail. In 1812 John Common designed a reaping machine, as did Patrick Bell another in 1828, but as long as there was an abundance of labour there was a reluctance to experiment with these machines. But from the 1850s the labour force began an irreversible decline, and the consequent shortage of labour and increase in wages revived interest in reaping machines. But it was the reaper designed by the American Cyrus McCormick which was adopted. In Britain in 1861 only 6 per cent of the grain harvest was cut by reaper, but by 1874 50 per cent and by 1900 80 per cent. In Western Europe the decline of the agricultural labour force came later and was slower, while farms were generally smaller than in England; thus the adoption of the reaper was slower. In 1882 only 1 per cent of the Netherlands grain was cut with a reaper, 6 per cent in France, 3 per cent in Germany, and even in Belgium, where the labour force had declined from the 1860s, only 4 per cent. The situation in the European areas overseas was quite different. Here the factor was not so much change in the numbers of the population but the very low density of population and the consequent shortage of labour. McCormick's reaper was patented in the 1830s. In 1851 less than 1 per cent of the cereal area of the United States was cut by reaper, but by 1870 four fifths.[9]

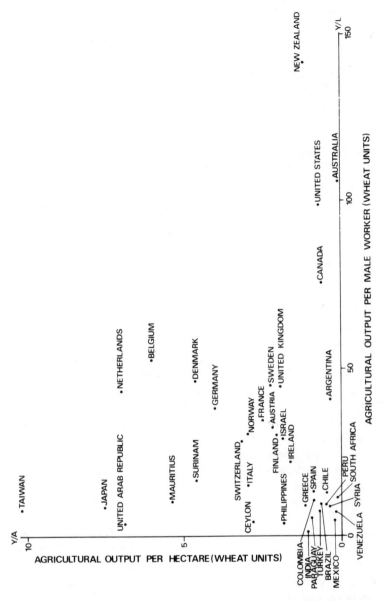

Figure 16 *Land productivity and labour productivity in selected countries, 1957–62*
Source: Y. Hayami and V. W. Ruttan, *Agricultural Development: An International Perspective* (Baltimore, 1971), p. 71.

Factor endowment and productivity

Not only do changes in the numbers of the agricultural population over time influence the adoption of new technologies, but differences in the man/land ratio in different countries may influence the type of new technology that is pursued, as just noted. Where land is in short supply and labour abundant there is little incentive to adopt labour-saving machinery, but every incentive to try and increase yield per hectare. On the other hand where land is abundant and labour scarce then machines are substituted for labour. This difference can be seen in Figure 16 where output per hectare and output per man-hour are plotted against each other for a number of countries in 1957–62. A group of countries which were sparsely populated – Canada, the United States, Australia and New Zealand – have low output per hectare but high output per man. In contrast countries with a high population density – Taiwan, Japan and Egypt – had high yields per hectare but a low output per man. Intermediate were the countries of Western Europe with densities between Asia and the European settlements overseas. These differences reflect the choice of inputs; this can be seen by plotting fertilizers against tractors (Figure 17). In 1960 countries with a large number of tractors per agricultural worker had low fertilizer inputs per hectare: Australia, New Zealand, Canada and the USA all fall into this category. Conversely countries with high fertilizer inputs and high yields had a small amount of tractor power per man. Japan, Taiwan and Mauritius fell into this class as did the Netherlands and Belgium. Most other West European countries fell into an intermediate category, with fairly high fertilizer and tractor inputs. It is thus possible to argue that the historical path of agricultural progress will be determined by the size of the labour force and the availability of land. Thus in the United States abundant land and a shortage of labour caused farmers to attempt to maximize output per man; fertilizers were neglected, but there were substantial investments in labour-saving machinery. Conversely in Japan a shortage of land led to a concentration upon land-saving inputs such as fertilizers and new rice varieties. Within Europe contrasting paths were taken by the Netherlands and Belgium, with high population densities and thus an emphasis on inputs that increased yield, and England with a relatively large supply of land per head of the labour force and thus relatively greater inputs in labour-saving machinery.[10]

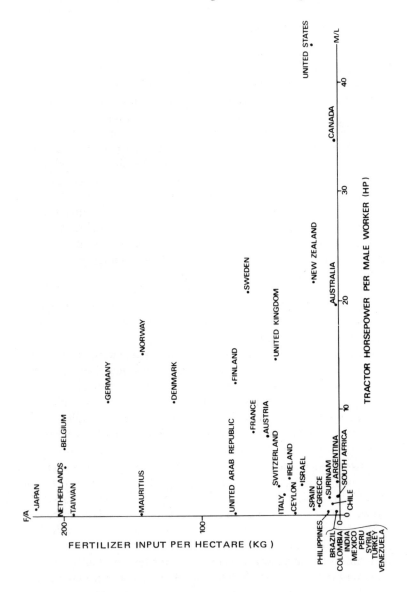

Figure 17 *Fertilizer inputs and tractor numbers in selected countries,*
1957–62
Source: Y. Hayami and V. W. Ruttan, *Agricultural Development:*
An International Perspective (Baltimore, 1971), p. 72.

Conclusions

The industrial revolution profoundly disturbed the balance of traditional agriculture; a primary consequence was the growth of a large commercial market which ended subsistence agriculture, and increased the need to move agricultural goods long distances. In many parts of Western Europe the slow increase in agricultural productivity was a restraint upon rapid industrialization, a restraint overcome in England by the import of food on a large scale. Because the early stages of economic development have also been periods of population growth there was little incentive to adopt labour-saving machinery until the labour force declined. But the adoption of such machinery is not a function only of changes in the labour force over time; differences in density may determine the type of progress that is possible.

9 Industrialization, demand and new technologies

Two further consequences of industrialization are considered in this chapter; the effects upon demand and the provision of new inputs. Both have had pronounced effects upon the type of agricultural goods produced, and the efficiency with which they are produced.

Income changes and agriculture

In most parts of the developing world today incomes are low and this is reflected in the diet of the majority of the population. Less than a tenth of the calorie intake is derived from animal foods compared with 30 per cent in the developed countries, and the average daily consumption is only 2282 calories compared with 3373 calories in the developed countries. Recent estimates suggest that about a fifth of the population of the developing countries have a diet below the minimum requirements necessary for health.[1] There is little or no reliable information upon average food consumption in the past but such as there is suggests that average food consumption in Western Europe before 1850 was similar to that in the developing world today. In France in the early nineteenth century average calorie consumption was 1700–2000 per day, with 80 per cent of the intake derived from bread; by 1850 72 per cent of the intake was from bread but daily consumption had risen to 2500–3000 calories. In Amsterdam in the 1850s the average daily intake was only 2267 calories, two thirds of which was derived from grain and potatoes. It does not follow that consumption levels were lower than this before the early nineteenth century. For most of the population consumption levels are likely to have followed the course of real wages which were inversely related to population growth. According to Wilhelm Abel they did not reach the high level attained in the fifteenth century until after 1850.[2]

There is much disagreement between historians as to whether

the industrial revolution brought any immediate benefits to the population. Studies of trends in English real wages have proved inconclusive; but they may have risen in the 1820s, and all are agreed that from the 1840s living standards were rising. This had two consequences. First, increased income could be spent on more food and more expensive foods; and second, a greater proportion of income could be spent on non-foodstuffs, thus creating an increased demand for industrial products, many of which in turn used agricultural products as raw materials. This was important. Bairoch has estimated that in pre-industrial societies 70–80 per cent of consumer incomes was spent on food; in rural India today 70 per cent is so spent. This proportion has continuously declined as real incomes have risen in Western Europe and the European settlements overseas. In the United States in 1869–78 35 per cent of income was spent upon food but this had fallen to 28 per cent in 1919–28. Comparable figures for Sweden are 39.5 per cent in 1871–5 and 29.3 per cent in 1926–30, and for the United Kingdom 26.5 per cent in 1900–10 and 23.7 per cent in 1960–5.[3]

The agricultural consequences of rising incomes

The first reaction to a rise in income in a poor society, such as much of Western Europe was before 1850, is to increase consumption of the staple products, cereals and potatoes, to provide an adequate calorific consumption. Once hunger is satisfied there is a shift to the preferred staple. In England and Wales in 1650 wheaten bread was eaten by less than half the population, the remainder relying upon rye bread, oatmeal or barley gruel. By 1750 62 per cent consumed wheaten bread, in 1800 70 per cent, in 1850 90 per cent and by 1900 wheaten bread was consumed by virtually all.[4] A further rise in incomes leads to a switch to more palatable but more expensive foods, generally with a higher protein content, including green vegetables, milk and meat. This may lead to an absolute decline in the consumption of bread and potatoes. The trends in consumption in the UK illustrate these changes, although the two world wars checked the decline in bread and potato consumption (Figure 18).

Until the end of the eighteenth century a large proportion of arable land in Western Europe was devoted to crops for direct human consumption, and little arable land was used for feeding livestock. In the nineteenth century not only were fodder crops

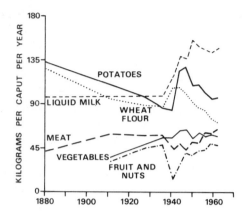

Figure 18   *Trends in food consumption in Great Britain, 1880–1970*
Source: J. C. McKenzie, 'Past dietary trends as an aid to prediction', in T. C. Barker, J. C. McKenzie and J. Yudkin (eds.), *Our Changing Fare: Two Hundred Years of British Food Habits* (London, 1966), pp. 135–49.

grown, but grain began to be grown for animal consumption, and was also imported to be fed to animals. This reflected the long term rise in incomes and increased demand for meat and milk. In 1815–24 livestock accounted for about 42 per cent of the value of British agricultural output; a more fundamental shift took place between the 1860s and 1930s (Table 13). There has not been any major shift in the structure of British agricultural output since; in 1970 69 per cent of British farm receipts were from livestock products, 18 per cent from farm crops and 11 per cent from

Table 13   *Gross agricultural output in the United Kingdom, 1867–9 and 1935–9*

	1867–9 (£ million)	Percentage of total	1935–9 (£ million)	Percentage of total
Crops	94.3	41.0	37.2	15.2
Horticulture	9.8	4.3	29.2	11.9
Livestock	125.7	54.7	178.1	72.9
Total	229.8	100.0	244.5	100.0

Source: E. Ojala. *Agriculture and Economic Progress* (Oxford. 1952). p. 208.

horticulture. A comparable shift from crops to livestock occurred elsewhere in Western Europe, and data for Sweden exemplify this trend (Table 14).[5]

While the shift towards livestock production was general throughout Europe in the second half of the nineteenth century, it was not achieved by the same means in each country. Denmark was already a major butter exporter in the 1850s, but the fall in cereal prices from the 1870s further increased specialization in livestock products, particularly dairying and pig production. This was achieved not by laying land down to grass, but by feeding cereals and roots to livestock, as well as importing feedstuffs. Livestock were stall-fed for much of the year and grass was a minor part of land use. Ireland presented a marked contrast. On the eve of the famine half the cultivated area was in arable. This continuously declined until by the 1960s grass occupied four fifths of the cultivated area. Agricultural output however is dominated by livestock products – store cattle for export to Britain, fat cattle and dairy products. In Britain the arable acreage declined and grassland increased, while the number of dairy cattle increased; but neither butter nor cheese were important products. Milk was largely consumed fresh in the cities.[6]

Although fruit and vegetable production has a long history, the second half of the nineteenth century saw a very rapid expansion of horticulture. One reason for this was improvements in transport, which made the long distance movement of perishable products possible. Another was the rise in income which made the purchase

Table 14 *Income from crops, livestock, horticulture and fishing, Sweden, 1861–1930*

	Crops		Livestock		Horticulture and fishing		Total
	Million kronar	*Percentage of total*	*Million kronar*	*Percentage of total*	*Million kronar*	*Percentage of total*	*Million kronar*
1861–65	97	47.1	92	44.6	17	8.2	206
1876–80	146	43.1	160	47.2	33	9.7	339
1901–05	139	29.6	279	59.4	51	10.9	469
1926–30	224	20.2	735	66.3	150	13.5	1109

Source: E. Ojala, *Agriculture and Economic Progress* (Oxford, 1952), Table 13.

of these for the most part expensive goods possible. Horticulture has not become of major importance in Britain; nonetheless the value of output rose from 4 per cent of total output in the 1860s to 11–12 per cent in the 1960s. Rising incomes in the United States and Western Europe prompted the expansion of fruit and vegetable production in these countries. But the fall in real transport costs also made it possible for areas at some distance but with a comparative advantage in climate to specialize in fruit or vegetable production. The banana trades provide an excellent instance of this. Essentially a tropical crop it needs to be shipped within a fortnight of cutting and kept cool during the voyage. The export of bananas from the Caribbean awaited the replacement of sail by steam, the spread of the railway and the introduction of refrigeration into ships. The first bananas reached Boston from Jamaica in 1870. In the 1880s American companies established plantations on the Caribbean coastlands of Costa Rica, Nicaragua and Panama. Later disease destroyed much of this and the industry was established upon the Pacific coastal areas of Central and South America.[7]

Industrial demand for agricultural products

The industrial revolution created a great demand for inorganic raw materials for industry, and indeed since the eighteenth century industry's dependence upon agriculture for its raw materials has greatly declined as metals, bricks and plastics have replaced timber, coal and oil have replaced wood as a fuel, synthetic fibres have supplemented wool and cotton, and chemical dyes have replaced woad and madder. But initially the industrial revolution created a great demand for established agricultural products, and later in the century new crops were in demand.

The textile industries were the pace-makers of the industrial revolution in Britain, Europe and the United States, and it was not only the rise of population but the rise of incomes that made this possible. Pre-industrial societies spent some 80 per cent of their disposable incomes upon food and 12–14 per cent on clothing. As real income rose and foodstuffs took a diminishing proportion of total income, so more could be spent upon clothes. The textile industry was the first to be mechanized and the first to create a mass demand for agricultural products. In the 1770s the British cotton textile industry obtained its supplies from the east Mediterranean and the West Indies. But the rapid growth of demand soon outran the capacity of these areas. In the southern United States however conditions were ripe for expansion. The plantation system, based upon slavery, already existed; the introduction of new cotton varieties made picking – a hand process – easier, and in 1793 Eli Whitney's improved cotton gin increased enormously the amount of cotton that could be separated in a day. Perhaps most important there was unsettled land to the west with climate and soil suitable for cotton growing. In 1790 United States cotton output was insignificant compared with that of India, Mexico or Brazil but by 1821 it had overtaken India, the leading producer, and by 1841 output exceeded that of the rest of the world combined. This was largely the result of a great increase in the area under cultivation. Confined mainly to the eastern seaboard in 1790, cotton planters had reached the lower Mississippi in the 1820s and were deep into Texas on the eve of the Civil War.[8]

The growth of the woollen textile industry had perhaps less dramatic consequences than the expansion of cotton textiles. Nonetheless the growth of the industry soon exceeded Europe's capacity to meet demand. However Australia received its first colonists in 1789 and in 1797 the *merino* sheep was introduced. With a great unoccupied interior, graziers could increase their numbers of sheep at an extraordinary rate, from 50,000 in 1813 to 21 million in 1860 and 100 million by the end of the century.[9]

The great growth of European industry in the nineteenth century increased the demand for a wide range of agricultural products, some new, some merely an expansion of existing products. Europe

Table 15   *The purchase of industrial inputs by farmers, Sweden*

	Agricultural workers as a percentage of all workers	Value of agricultural output as a percentage of gross domestic product	Value of purchased inputs as a percentage of value of agricultural production
1860	64.0	31.2	5.5
1890	59.2	25.8	12.0
1940	28.8	12.8	22.8
1960	13.8	7.3	39.7

Source: A. Simantov, 'The dynamics of growth of agriculture', *Zeitschrift für Nationalökonomie*, vol. 27 (1967), pp. 328–51.

derived its vegetable oils from olive oil in the south and rape and flax in the north. The demand for oil as a lubricant, for soap, varnishes, and as a food prompted the expansion of a variety of tropical crops in the second half of the nineteenth century. Later the electrical industry and then the motor vehicle industry created a demand for rubber. Indigenous to the Americas, the early demand for rubber was met by the tapping of scattered trees in the Amazonian rainforest, but later the establishment of plantations in Malaya and elsewhere in south east Asia transformed the economies of those countries. In the twentieth century the development of synthetic rubber has inhibited the further expansion of rubber plantations, a consequence of later industrialization that has also affected the cotton and woollen producers.[10]

Industrialization and new technologies on the farm

Industrialization created new demands for agriculture; it also created new inputs and improved established methods. Combined with the spread of scientific research in agronomy, industrialization has transformed the productivity of modern agriculture; but while the antecedents of these new inputs can be traced back a century and a half, it is only in the last forty years that they have been widely adopted in western agriculture.

Traditional farmers bought little off the farm (see pp. 97–8), but in every country that has industrialized the value of inputs purchased off the farm has increased, as the case of Sweden demonstrates (Table 15). In the United States in 1940 the value of inputs purchased off the farm was 34 per cent of the value of all farm inputs, but this had reached 62 per cent by 1960.[11] The reasons for this remarkable growth have been twofold. First, advances in the techniques, organization and scale of industry have greatly reduced the real cost of purchased inputs. Thus the cost of superphosphate in England fell from £8 per tonne in 1858 to £3 per tonne in 1934.[12] Second, industrial inputs have a greater capacity than traditional inputs to increase productivity. This is best exemplified by the use of machinery. In 1800 in the USA it took 373 man-hours to produce 2700 kg of wheat using an ox-drawn plough, the sickle and the flail. By the 1960s using tractors and combine harvesters it took only ten hours. Crop yields in Western Europe and the United States are now three to four times what they were in 1800.[13]

Land-saving inputs

Farm inputs can be usefully divided into those that increase yield per unit area, or land-saving, or those that increase output per man hour, or labour-saving. They are not always mutually exclusive in their results. Thus the use of machinery to harvest may rescue a crop at a time of uncertain weather and so increase yields as well as saving labour. Before the nineteenth century few innovations were overtly labour-saving, most aimed at increasing output per hectare.

Improved crop varieties

The selection of seed had doubtless long been going on before the nineteenth century, but there was little knowledge of the genetics of crop improvement. A principal means of increasing yields in this way was to import crop varieties that had done well elsewhere in similar climates; so in the nineteenth century hard wheat varieties were taken from the semi-arid parts of southern Russia to the American west as they had earlier been introduced into southern France. The rediscovery of Mendel's work on genetics allowed plant breeders to select for specific characteristics such as drought resistance, response to fertilizer, a short straw, or immunity to specific diseases. Crop breeding has been responsible for a considerable part of the increase in yields in the last forty or fifty years. In Western Europe it is thought to account for 25 per cent of the yield increase of cereals between 1940 and 1970, and in the United States 48 per cent of the yield increase of wheat between 1935 and 1975. But crop breeding has not only increased yields; it has also helped extend cultivation into areas which were too dry or had too short a growing season[14] (see p. 67).

The reduction of plant and animal disease

It has been estimated that a fifth of the world's grain output is lost to pests and disease every year, while an unknown but considerable proportion of the standing crop is destroyed before harvest by pests and disease. The pre-industrial farmer had little protection against either; fallowing prevented the build-up of soil-borne disease, while seed that appeared immune could be selected. The loss by pests and disease has been neglected by historians, except in the more dramatic cases such as the potato blight in Ireland in 1845 or the *phylloxera* outbreak which began in French vineyards in 1863; such losses were probably as serious as losses from bad weather. The hazards of these pests have been greatly reduced in modern times by the development of chemical pesticides and fungicides. These may be said to have begun in France in 1851 when it was found that vines could be protected against mildew with a mixture of sulphur and lime, and the later discovery in England that copper sulphate was effective in the same task. A number of chemical pesticides were developed in England before the First World War, and aerial spraying was practised in the United States in the 1920s. The discovery of DDT in Switzerland just before the Second World War was a major breakthrough. It is however only since 1945 that pesticides and fungicides have been widely used in western agriculture.[15]

The preparation of land and the elimination of weeds

The traditional farmer spent much of his time on the preparation of the seed-bed and the elimination of weeds. Both helped to increase yields. Before 1800 slow advances in the design of ploughs and harrows had contributed to the better creation of a seed-bed, but for the elimination of weeds the farmer was largely dependent upon the cultivation of the fallow and weeding by hand with hoes. Jethro Tull's development of the seed-drill and the horse-hoe in the early eighteenth century was thus of great importance, for it allowed inter-row cultivation during plant growth which had been impossible when crops were broadcast. But the adoption of the drill and the horse-hoe was remarkably slow. In Norfolk in 1800, the classical centre of the English agricultural revolution, the drill was hardly used; inter-row cultivation of root crops was common in the nineteenth century but not of grain. In France in 1862 only 7 per cent even of large holdings used drills and they were little used in the United States before the Civil War.[16]

Nonetheless the industrial revolution had two important consequences for the preparation of the seed-bed. First, the fall in the cost of iron meant that implements were increasingly made from iron and later steel, rather than wood with only the cutting edges made from metal. Second, implements and machines were increasingly made in factories rather than by local craftsmen, reducing their cost and increasing their efficiency. In eastern

England agricultural machinery firms became widely established after the end of the Napoleonic Wars, and they produced a variety of implements which helped in the cultivation of the land; not only ploughs but rollers, horse-hoes, clod-crushers and harrows were soon available. Such advances did not relieve hand labour from the tasks of weeding; root crops were lifted by hand until after the Second World War in England. Here again the chemical industry came to the aid of the farmer. In the 1930s a number of chemical weed-killers were developed in Britain and Europe of which 2,4–D was the first of commercial importance. The use of herbicides, which has been largely confined to the postwar period, has greatly reduced the amount of labour needed in crop production.

As has been seen earlier (pp. 61–5) the improvement of cultivation practices was particularly difficult upon heavy clay soils which were hard to till and difficult to drain. Until the seventeenth century surface drainage in the form of ridge and furrow was the only means of removing water, but in the seventeenth century 'hollow' drainage was slowly adopted; stones or thorn bushes were placed in trenches and covered with soil. Little is known of the efficiency or extent of this method of drainage, but it was replaced in the early nineteenth century by the use of horse-shoe shaped tiles and later by pipes, while the mole-plough speeded the process of hollow drainage. In 1846 a Public Money Drainage Act provided government finance in England, but it is not clear what proportion of the land in need of underdrainage was treated; by 1880, of the land in need of drainage, probably no more than one fifth had been drained. In France underdrainage was recorded in the agricultural statistics from 1852, but less than 1 per cent of the cultivated area had been drained by the 1890s, mainly in the Paris basin. Tile drainage lasted no more than thirty years; in the years of depression in the late nineteenth century and the interwar period there was little extension of the area drained, and much old drainage was not renewed. Since 1940 however there has been a great revival in underdrainage in England and Wales.[17]

The use of artificial fertilizers

Although farmers before 1800 used a wide range of substances as fertilizers, livestock manure was the most important source. The increasing livestock density carried by the new mixed farming increased the supply of manure, and the import of oil-seeds from the 1820s to make oil-cake to feed the cattle increased the supply and the quality of the dung. In 1841 J. B. Lawes established a factory to produce superphosphate at Deptford and by 1870 there were eighty such factories in Britain. Superphosphate was made by dissolving bones in sulphuric acid but later coprolites and then Norwegian apatite were used. Guano was imported in large

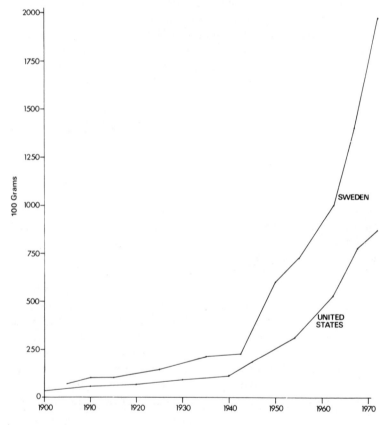

Figure 19   *Consumption of artificial fertilizers in the United States and Sweden, 1900–70: $N + P_2O_5 + K_2O$ per hectare of arable land Sources*: FAO, *Fertilizer Yearbook, 1978* (Rome, 1979); H. Osvald, *Swedish Agriculture* (Stockholm, 1952), p. 52; US Department of Commerce, *Historical Statistics of the United States: Colonial Times to 1957* (Washington DC, 1960), p. 285.

quantities between the 1830s and 1870s. In the second half of the nineteenth century ammonium sulphate – a by-product of gas production – was produced as a nitrogen fertilizer, although it was not widely used. In the 1860s Britain used more artificial fertilizers than the rest of Europe combined; indeed superphosphate was not manufactured in Europe until 1880. In the later nineteenth century basic slag, a by-product from the treatment of phosphoric iron ores, was used on acid grasslands. The next major advance came in Germany where Fritz Haber discovered how to fix nitrogen synthetically; in the 1930s compound fertilizers, containing a combination of nitrogen, phosphorus and potassium, were being produced.[18]

The consumption of artificial fertilizers increased rapidly in Britain, Western Europe and the United States in the second half of the nineteenth century. Thus US consumption of commercial fertilizer rose from 48,000 tonnes in 1850 to 281,000 tonnes in 1870 and in Britain from 51,000 tonnes in 1845 to 234,000 tonnes in 1916. The rate of increase then slowed down and usage was about 305,000 tonnes in 1938; but this meant there was little used per hectare. In the interwar period only potatoes and sugar-beet received significant quantities, and hardly any fertilizer was used on grass in England and Wales. By the 1920s many parts of north west Europe were using greater amounts of fertilizer per hectare than England and Wales. However since 1945 there has been a dramatic increase in fertilizer use in Britain as there has indeed been in the United States and Western Europe (Figure 19). In the United States the use of fertilizers quadrupled between 1940 and 1966. In France the consumption of nitrogen fertilizers sextupled between 1950 and 1974 while the use of phosphorus fertilizers more than tripled. Over the same period in Germany the consumption of phosphorus fertilizers doubled and that of nitrogen fertilizers nearly quadrupled.[19]

Industrial inputs and crop productivity

Although industrial growth began to produce new inputs for agriculture from the mid nineteenth century their adoption was slow, and most yield increases before the 1880s must have been due to the wider use of pre-industrial techniques, such as the increased growth of legumes. Between the 1880s and the 1930s there were divergent trends between crop yields in Western Europe

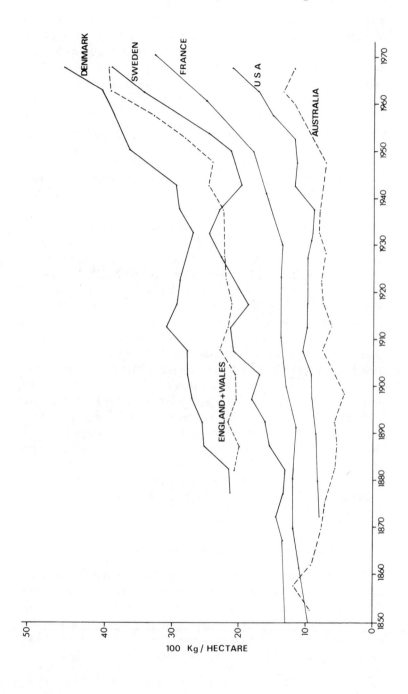

and the European settled areas overseas. In the latter, with abundant land but a shortage of labour, little effort was made to increase crop yields; average wheat yields in Canada, the United States, Australia and Argentina remained at about 670 kg/ha in this period. In Western Europe there was a gradual increase in the use of artificial fertilizers and improvements in other farming practices, and wheat yields rose by about one third. But since the 1930s there has been a great increase in the application of fertilizers, made possible by the falling real cost of fertilizer, and particularly nitrogen fertilizer, much improvement in crop varieties, and the widespread application of pesticides, fungicides and herbicides. Consequently wheat yields have more than doubled in Western Europe and also in the overseas European settled areas (Figure 20). Thus although the antecedents of industrial inputs can be traced back to the mid nineteenth century, the benefits were limited and confined to Western Europe until the period after the Second World War.[20]

Labour-saving inputs

The nineteenth century was the age of steam, yet the new source of power had little effect upon agriculture even in Britain or America; it was used to drain polders and fens from the 1820s, and applied ingeniously but not very successfully to ploughing. It was only in threshing that steam became widely used, mainly after the

Figure 20   *Crop yields in selected countries, 1850–1970*
             Sources: W. Malenbaum, *The World Wheat Economy, 1885–1939* (Cambridge, Mass., 1953), pp. 236–9; FAO, *Production Yearbooks 1950–72* (Rome, 1951–73); Ministry of Agriculture, *A Century of Agricultural Statistics: Great Britain 1866–1966* (London, 1968), pp. 108–9; E. Jensen, *Danish Agriculture* (Copenhagen, 1937), pp. 370–92; H. Osvald, *Swedish Agriculture* (Stockholm, 1952), p. 50; US Department of Commerce, *Historical Statistics of the United States: Colonial Times to 1957* (Washington DC, 1960), pp. 297–8; P. M. Hohenburg, 'Maize in French agriculture', *Journal of European Economic History*, vol. 6 (1977), pp. 63–101; E. Dunsdorfs, *The Australian Wheat-growing Industry 1788–1948* (Melbourne, 1956), pp. 534–5; G. Sundbarg (ed.), *Sweden: Its People and Industry: A Historical and Statistical Handbook* (Stockholm, 1904), pp. 52–5, 528.

1840s in Britain; it replaced the flail, or the horse-driven threshing machine, which had been invented by Andrew Meikle in Scotland in 1786. (This itself had effected considerable economies: a man with a horse-driven thresher could thresh five to six times as much grain as a man with a flail.) But otherwise steam had little effect on agriculture; indeed the major change of power in the nineteenth century was the completion of a process that had begun in the ninth or tenth century: the replacement of the ox by the horse. The latter only overtook the ox in the Netherlands after 1800, and in the United States in the 1840s. This undoubtedly increased productivity; the horse was faster than the ox although it had no greater traction power. In the seventeenth century oxen could plough 0.4 ha a day, the horse 0.5 to 0.6 ha, but with the improved lighter ploughs that became available in the eighteenth century a horse could manage 0.8 ha in a day.[21]

The triumph of the horse was shortlived. By the end of the nineteenth century the ox was a historical curiosity in England. Experimental petrol-driven tractors were developed in the USA in the 1890s and the early twentieth century; they increased in usage in the 1920s on farms in Britain and the United States. But even in the 1940s the horse was still providing half the traction power on farms in both countries. Since then their demise has been rapid (Figure 21). In an age of cheap oil tractors were cheaper and quicker; further, the decline of the horse released the land needed to feed it for more productive purposes. In 1914 the land used to feed horses accounted for 20 per cent of the gross agricultural output in the United States and 17 per cent of the man-hours. Nor was the petrol-driven engine the only new source of power. In the 1930s only 10 per cent of farms in the United States had electricity. By 1960 only 3 per cent were without.[22]

The changing form of power on the farm was a major source of productivity growth on farms, but industrialization had little role to play in this – other than in steam ploughing and threshing – until the invention of the tractor and its mass production in the

Figure 21 *Horses and tractors in Great Britain and the United States, 1910–70, per 400 ha of cropland*
Sources: Ministry of Agriculture, *A Century of Agricultural Statistics: Great Britain, 1866–1966* (London, 1968), pp. 73, 129; US Department of Commerce, *Historical Statistics of the United States: Colonial Times to 1957* (Washington DC, 1960), pp. 285, 290.

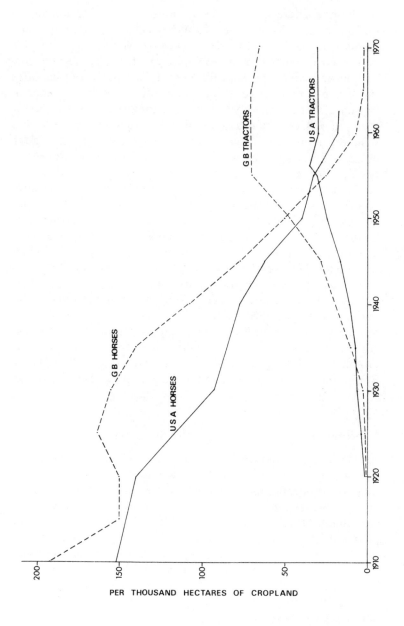

PER THOUSAND HECTARES OF CROPLAND

twentieth century. It was different with farm implements and machinery. There was of course a long history of technological change in agriculture before the industrial revolution, but implements were mainly hand-made, from wood, with iron used only for cutting edges. The advances in the iron industry reduced the cost of iron and later steel, and from the 1820s specialized factories producing a wide range of agricultural implements appeared in Britain and the United States. The cutting of the grain harvest received most attention, and the sickle was replaced by the scythe, then the reaper and finally the combine harvester; the latter developed in California in the 1880s and was then horse-drawn, but later became tractor-driven and finally was self-propelled. This has led to astonishing advances in labour productivity. In the early nineteenth century one man with oxen, sickle and flail could cultivate no more than 3 ha of wheat a year. In the 1890s one man, before the use of the tractor or the combine harvester, could farm 55 ha and advances since have been even greater. The time taken to produce 2700 kg of wheat in the United States fell from 383 man-hours in 1800 to ten man-hours in the 1960s.[23]

Attempts were made to mechanize other farm activities in the mid nineteenth century; there were even potato pickers. Parts of hay-making were successfully mechanized, but on the whole few advances were made with labour-saving devices in livestock production. Milking machines were invented in the 1890s but in Britain they only replaced hand-milking after the Second World War.[24]

Conclusions

Industrialization had an immediate effect upon agriculture not only in the countries where manufacturing industry became established but throughout the world. Demand for raw materials created new agricultural regions in the American South, in Australia and South Africa. Rising incomes shifted West European farms progressively towards livestock production, but also created new livestock regions such as the pampas of Argentina and the dairy farms of New Zealand. In contrast the creation and adoption of new industrial technologies was longer delayed. Between the 1850s and the 1930s the adoption of machines and fertilizers was slow; but since then Western agriculture has been revolutionized by new industrial technologies.

10 Transport and agricultural change

The importance of transport costs in determining geographical variations in the type of farming is widely acknowledged, while changes in transport costs over time have also had a profound impact upon agriculture. In the economic model put forward in 1826 J. H. von Thünen argued that the cost of transporting agricultural produce to market was the major determinant both of the intensity with which a crop was grown, and the combination of crops that a farmer would grow.[1] Von Thünen wrote at a time when industrialization had relatively little impact upon European agriculture, and his model helps to explain some of the features of the agricultural geography of pre-industrial Europe. He did not consider the dynamic aspects of his model – the extent to which *changes* in transport costs affect intensity and land use patterns – but later writers have done. Nor could he foresee the very important subsequent changes that there would be in transport: not only did real costs fall, but it became possible to move perishable commodities much greater distances. In this chapter von Thünen's model is briefly outlined; its usefulness in interpreting pre-industrial agriculture is discussed; and then the long term fall in transport costs after von Thünen's time is described, and the consequences for agriculture are assessed.

Von Thünen's model has been widely discussed and only an outline is given here. He attempted to analyse the influence of distance from the market upon the farmer's choice of crops and the intensity with which they were grown; to demonstrate the importance of distance he constructed an *Isolated State*, an economic model where all other variables which might influence a farmer are held constant. Farmers were assumed to maximize their profits, had access to recent information upon prices, and were prepared to change their methods of farming accordingly. The Isolated State was a plain with no differences in soil or climate, with a radius from a central town, where all produce was sold, of

370 km. Beyond this was a wilderness; there were no imports or exports. All produce was carried to market by waggon, transport costs were proportional to distance, and every farmer could take the most direct route to market. Farmers paid for transport to market, and all prices were set in the single central market.

Using a mixture of deductive concepts and actual production costs of the time, von Thünen then outlined two theories. In his *intensity theory* he showed that if only one crop was grown farmers would maximize their profits – although he used the concept of economic rent, not profit – if they farmed intensively near the market, and more extensively as distance from the market increased. By intensity he meant labour inputs per unit area. Although there have been few explorations of this idea, it is confirmed by modern economic theory. Farmers who wish to maximize profit will increase their inputs until the cost of the marginal input equals marginal revenue. As prices rise going towards the town inputs will thus be increased; as the price received by the farmer falls with increasing distance from the town, inputs will be reduced.

Von Thünen's *crop theory* is much better known, but its rationale is less clear. He argued that with increasing distance from the town it would pay the farmer to replace one crop with another. The logic of this has been stated in modern terms by A. Lösch and E. S. Dunn,[2] using the formula:

$$R = E(p-a) - EfK$$

where R is the rent per unit of land, K is distance, E is the yield per unit of land, P is the market price per unit of commodity, a is the production cost per unit of commodity, and f is the transport rate per unit of distance for each commodity.

They argued that the crop that produces the greatest economic rent per unit area will occupy sites nearest the town; crops that yield a lower economic rent will be displaced to greater distances. Thus three different crops A, B, and C all give a positive economic rent at the market, A the highest, C the lowest (Figure 22). But the cost of transporting crop A to market is high, and the rent gradient falls rapidly to zero at MD. Crop B has a lower economic rent per hectare than A but a less steep rent gradient. To maximize profits farmers between M and J_1 will grow A rather than B, although both give an economic rent. But beyond J_1 crop B gives a higher economic rent than A, because of its shallower rent gradient, and becomes the preferred crop. A third crop, C, yields a lower rent

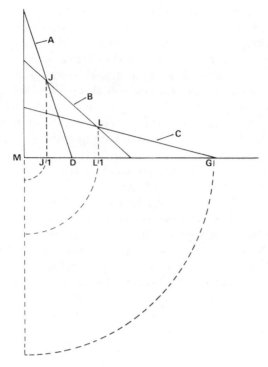

Figure 22 *Economic rent and crop zones*

per hectare at the market, but because transport costs are a lower proportion of the value of the crop, the rent gradient is shallower, and beyond L_1 crop C gives a higher economic rent than either crop B or crop A. At G, crop C gives a zero rent per hectare, and assuming only three crops are grown, then this is the extensive margin of cultivation. Arcs drawn from J_1, L_1 and G show how concentric crop zones will develop at varying distances from the market.

Von Thünen's own exposition of his crop theory is less simple than this modern restatement. He tried to show how various crops and combinations of crops would succeed each other with increasing distance from the market, by calculating economic rent for each product from the accounts of his own farm. The resultant pattern is shown in Figure 23 and Table 16. In the Free Cash Cropping zone perishability is a major determinant of nearness to market, while the outer limit was decided by the distance that manure from the town could be transported. The zone of Forestry was near to the town because the low value per unit weight made it impossible to transport firewood or timber any great distance, yet there was a great demand for wood in pre-industrial cities. But forestry is extensive, not intensive. Beyond the Forestry zone lay three mixed farming systems that differed in the amount of fallow that occurred in the rotation and hence the labour inputs per hectare. The Crop Alternation System had no fallow, was the most intensive, and was nearest to the market; the Three Field System with most fallow, and the least intensive of the mixed farming systems, was the most distant.

Beyond the mixed farming systems lay a zone of Stock Farming in which were reared young cattle to be driven to market and fattened on pastures near the city, and also sheep for wool; but it also included dairy cattle whose milk was converted into butter, grain that was converted into alcohol, and highly labour-intensive crops such as flax and tobacco that were processed before being marketed. Thus intensity is not the only factor determining the land use zones; perishability accounts for the location of vegetables and fresh milk near the market; the very high transport cost per unit weight of potatoes, timber and hay ensures they are grown near the market. Conversely if milk is converted into the less

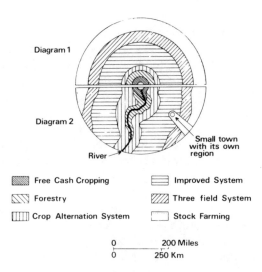

Figure 23   *Von Thünen's type of farming zones*
            Source: P. Hall (ed.), *Von Thünen's Isolated State* (London, 1966), p. 216.

perishable and higher value products of butter and cheese they can be produced at remoter situations, as can flax and tobacco. Two final points can be made. First, von Thünen thought that his analysis was applicable to all scales; his Isolated State was 370 km in radius, but he believed that on a farm the distance between farm house and fields had similar affects upon intensity and crop selection. Later writers have applied the analysis to a variety of scales, from the parish to the world.

Second, von Thünen was well aware that the variables that he held constant in the model could be important in determining the location of agricultural production. He considered the influence of tariffs upon intensity, and briefly considered the consequences of multiple markets. Of more importance he allowed that a cheaper form of transport such as a river or canal would extend the zones

Table 16 *Land use zones in the Isolated State*

Zone		Distance (km)	Products
1	Free Cash Cropping	0–29.7	Fresh fruits and vegetables, milk, hay and straw, potatoes, cabbages, roots, fattening cattle on pastures. No fallow, no fixed rotation. Uses town manure
2	Forestry	29.7–52	Firewood, timber
3	Crop Alternation System	52–69.7	Potatoes, rye and root crop, oats, clover, wheat and a root crop *or* potatoes, barley, clover, rye, tares, rye
4	Improved System	69.7–183.3	Rye, barley, oats, pasture (three years) fallow. Cattle and sheep
5	Three Field System	183.3–233.8	Rye, barley, fallow. Cattle and sheep
6	Stock Farming	233.8–370.8	Butter, store cattle, sheep for wool, grain for alcohol. Rape, tobacco, flax
7	Wilderness	370.8	Forests: hunting for hides.

Source: W. B. Morgan. 'The doctrine of the rings'. *Geography*. vol. 58 (1973). pp. 504–5.

outwards; and that soils were unlikely to be of equal fertility. He thought that with uniform labour inputs on all soils, fertile soils would yield a greater economic rent than poor soils.

Transport costs in pre-industrial agriculture

Historians are agreed that the transport of agricultural produce before the nineteenth century was slow and costly; there are unfortunately no time series to demonstrate long term trends in the cost of movement, or comprehensive data to show the relative costs of different types of transport, or the relative costs of transporting different types of agricultural produce.

Overland costs were high because of the very poor surfaces of roads. Until the fourteenth century the pack-horse or mule was almost the only means of transport, and they could only carry loads of 100–150 kg. Four wheeled waggons increased the amount that could be carried per animal, but although they were known in the late Middle Ages they were still rare in Tudor England, and as late as 1637 there were none at all in Cornwall. But the use of waggons increased from the mid seventeenth century and by the early eighteenth century half the carrier services to London from the provinces were by pack-horse, half by waggon. Elsewhere in Europe the waggon was slow to replace the pack-horse. In Castile in the late eighteenth century 90 per cent of the transport of goods was still by pack-horse or mule.[3]

Transport by water was invariably cheaper, but until the middle of the seventeenth century there were few canals or navigable rivers anywhere in England, and only the larger rivers in Europe were so used. Thus coastal areas were particularly favoured. In eighteenth-century England the average cost of transporting goods by road was 8d. per tonne-kilometre, by river 1½d., while sea transport was one eighth of land transport. The ease of moving agricultural goods depended upon the value per unit weight of the produce. In the fifteenth century it cost only 1.5–2.0 per cent of the price of wool or wine to ship it by sea a given distance, but 15 per cent of the value of wheat. The grains differed among themselves. In the eighteenth century wheat could be shipped from north east Kent to London by water for 3–5 per cent of its price, oats 8 per cent and beans 10 per cent. Until the advent of the railway cattle were driven on the hoof to market; milk could not be marketed fresh more than 24 km from the dairy, although it could be converted to butter or cheese, which could be preserved longer.[4]

The village and the fields

Numerous modern studies have shown that the distance that has to be moved to and from the buildings of a farm to work in the fields – or from a village to outlying fields – influences the intensity of cultivation.[5] Does this throw any light upon pre-industrial agriculture?

From the twelfth to the nineteenth centuries the bulk of farmers in Western Europe lived not in isolated dwellings or hamlets but in villages, and walked out each day to work in the arable fields surrounding the villages. There was a marked zonation of land use in the open fields. Near the village were closes of grass for livestock, and gardens of intensive cultivation. Surrounding this were the great open arable fields, which were divided into long, narrow strips: each farmer's strips were intermingled with other strips rather than in contiguous blocks. Surrounding the arable fields was common grazing land and woodland. It was distance rather than soil type that determined this decline in intensity away from the village, with one exception. Meadowland, grassland flooded part of the year by the overflow of the river, was highly valued. This decline of intensity with distance was particularly marked in the infield–outfield system which survived in Ireland, Scotland, Norway, Sweden and other parts of Western Europe until the nineteenth century. In Scotland in the late eighteenth century the village arable was divided into two parts. The infield near the village was in tillage every year, and was heavily manured. The outfield was in two parts; the inner part, one third of the outfield, was in oats for 3–4 years followed by 5–6 years in grass with manuring in the first year. The remaining and more distant part of the outfield was in crops for 4–5 years followed by grass for 5–6 years. No manure was applied.[6]

In both the open field and the infield–outfield system labour inputs per hectare declined with increasing distance from the farm or village. Beyond a certain distance the cost of the time taken to move to and from the village exceeded the returns from the land cultivated; the extensive margin of cultivation had been reached. In sparsely populated areas then the limits to cultivation may be determined not by soil fertility, or even demand, but by distance.

One response to this is the growth of temporary settlements, which may later become separate and distinct communities. Such daughter settlements grew up between primary settlements throughout Western Europe in the twelfth and thirteenth centuries.

Land use patterns in the pre-industrial British Isles

The growth of the population of London, the richest market in England, illustrates the importance of distance upon land use in pre-industrial Britain. In the mid sixteenth century London drew its grain supplies not from the intermediate hinterland, but from Berkshire and Oxfordshire via the upper Thames, and from north east Kent. There were however already signs of more specialized farming; market gardening was introduced by Huguenot refugees in the mid sixteenth century, while hops were grown in Kent and Essex. As the city expanded the nearby counties were unable to satisfy the demand for grain, and by the later seventeenth century grain was coming from East Anglia, brought by coastwise shipping, not by land. In years of dearth supplies of grain were brought from the north east and south west, all by sea. More striking was the source of meat, hides and tallow. Cattle were first driven from Wales to London in the fourteenth century, but grew steadily in importance from the 1520s: in the seventeenth century young cattle from Ireland, Scotland and the north of England were being driven to the south east and fattened on good pastures before being slaughtered at Smithfield. By the 1650s about one third of the annual slaughter of livestock in England was of cattle reared in the Celtic countries. Little is known of London's milk supply at these early times. But by the end of the eighteenth century milk came from two sources. Cows were kept in the city and stall-fed with hay, roots and brewers' grains. But around the city, and particularly to the north, the predominant form of land use was permanent grass and meadow, which fed cows which provided milk for the city. About one fifth of the 620 km^2 around the city of London which was surveyed by Thomas Milne in 1800 was devoted to market gardening.[7]

Thus, taking London as the market, there was a crude zonation of land use around the city, with milk and vegetable production nearby, and livestock being drawn from the remoter areas, which of course were also those with upland environments unsuited to crop cultivation. In between lay the great area of lowland England devoted to mixed farming. It is not possible to discern any decline in the intensity of arable farming as distance from the metropolis increased; there were after all important provincial markets. It is true that the south east of England was an area where the open field system was earliest eliminated; but it is also true that the remoter areas lost their open fields at an early date[8] (see pp. 200–1).

Land use rings in Europe

The presence of market gardening and milk production near the big cities was as noticeable in Europe as it was around London. Wilhelm Abel has argued that a von Thünen zonation had appeared in Europe as early as the late sixteenth century.[9] The Low Countries were by then highly urbanized with a precocious industrial growth. In both Holland and Flanders this led to an early suppression of the open fields, and the decline of fallow; in Holland there was specialization in dairying, and in both Holland and Flanders industrial crops were grown, while the beginnings of a later major market gardening industry were apparent in the sixteenth century and boomed in the seventeenth century. Grain gave too low an economic rent to compete with these products; it was instead imported from the areas around the southern Baltic, where the three field system prevailed and yields were low. Cattle also moved into the Low Countries for fattening from Jutland. The urban areas of the Rhineland and other parts of south Germany acquired young cattle from Hungary, Bohemia and even Russia. The zone of forestry so conspicuous in von Thünen's model was absent from the Low Countries; timber was imported by sea from the Baltic countries, which along with Scandinavia were later to provide England with much of its timber.[10]

The decline of transport costs and changes in land use

Von Thünen's followers in the nineteenth century argued that increasing population and rising prices would lead to the expansion of intensive farming systems and the outward displacement of the more extensive systems. In this century August Lösch, Theodore Brinkmann and E. S. Dunn have also speculated on the consequences of economic change for the model.[11] Increasing population and rising prices would lead to an expansion of the margin of cultivation; each of the interior zones would also expand outwards. Rising per caput income would lead to a higher demand for meat and thus an outward expansion of the stock rearing zone. A decline in the real cost of transport would also lead to an outward expansion of the zones. Brinkmann argued however that if transport costs fell markedly then the advantages of location near to market would be less important and local advantages of soil or climate would become more significant in determining location. The greater speed of transport and the preservation of perishable commodities would also free products such as milk from their market orientation.

The fall in transport costs and improvements in food preservation

In the eighteenth and nineteenth centuries the cost of moving agricultural produce fell partly because existing modes of transport were improved – such as the better surfacing of roads – but also because a new mode of transport – the railway – was introduced. The fall in the real cost of transport continued into the twentieth century as the lorry began to displace the steam locomotive, and for some very high value products the aeroplane became important.

The cost of moving agricultural produce was falling before the industrial revolution. In England the turnpiking of roads and the more general use of the four wheeled waggon led to a continuous fall in the real cost of moving agricultural goods throughout the eighteenth century; the spread of the canal was also important for the cost by barge was half that by land. Maritime costs also fell in the eighteenth century, without any great increase in the capacity or speed of ships. The decline in piracy reduced insurance rates and the need for merchant ships to carry armaments; better organization meant less time idle in ports, and a better knowledge of currents and winds meant faster voyages. The cost of maritime transport fell at 0.5 per cent to 1.0 per cent per annum between 1650 and 1770, but warfare then halted this decline until the 1820s.[12]

The nineteenth century saw a more rapid decline of freights both on land and sea. Maritime freights fell an average of 3.5 per cent per annum from 1814 to 1860. On land the spread of the railway after the 1820s had spectacular effects. In Argentina freights by rail were one twelfth those by ox-drawn cart, in England the railway cut the cost of moving cattle in half, and in

South Wales the cost of moving butter by rail was one ninth what it had been by waggon. The shift from waggon to rail had dramatic consequences, but the increasing efficiency of the railway led to further declines. The cost of moving 0.35 hl of wheat from Chicago to New York fell from 34 cents in 1873–5 to only 8 cents in 1905. The spread of the railway revolutionized land use in remoter areas. By 1890 three quarters of the agricultural land in the United States was within 64 km of a railway line. But it was the later nineteenth and early twentieth centuries that saw the most spectacular decline in freights, particularly by sea as steam-driven steel ships replaced the wooden sailing boat. The cost of moving 3 hl of wheat from southern Russia to Britain fell from 8s.6d. in 1872 to 2s.3d. in 1900, from the United States to Britain from 6s.6d. to 2s.2d.; while between 1873 and 1896 the cost of bringing wool from Australia to Britain halved.[13]

Not only did costs fall in the nineteenth century, but advances in the preservation of perishable commodities greatly increased the distance they could be moved. Canning dates from the early nineteenth century, but it was the adaptation of refrigeration, first used in a Sydney brewery in 1851, to ships that transformed the transport of meat and butter. Frozen meat first reached Europe from the United States in 1874, from Argentina in 1877, from Australia in 1879 and New Zealand in 1882. Improvements in keeping milk cool on railways meant that by the 1870s milk could be delivered fresh as far as 240 km. Agricultural produce was increasingly processed in factories, increasing not only preservation but also the value per unit weight and allowing commodities to stand long distance journeys.[14]

London's milk supply

The fall in costs and the improvements in the preservation of perishable goods transformed the pattern of agriculture in the nineteenth century. Two examples will suffice to illustrate these changes: the provision of milk to London, and the location of zones of crop production in the United States.

In the middle of the nineteenth century London got most of its milk supply either from dairymen who kept cows in the city, or from dairy farmers on the edge of the built-up area (Table 17). By 1910 this was transformed. The consumption of milk rose sevenfold, little milk was still produced in the city, and the railway provided

Table 17   *Sources of London's milk supply (million litres)*

	Town milk	Road milk	Railway	Condensed	Total
1850	55	9–14	4	–	68–73
1870	40	10–20	33–50	1	84–111
1890	31–6	0–2	177–87	21	229–46
1910	12	0–1	304–409	45–91	361–513

Source: P. J. Atkins. 'The growth of London's railway milk trade. *c*. 1845–1914'. *Journal of Transport History*. vol. 4 (1977–8). pp. 208–26.

nearly all the fresh milk; in the 1920s this milk was brought from as far away as Derbyshire and Devon. Thus milk ceased to be produced in any quantity in or near London, and was brought from those areas that had a comparative advantage in the production of grass. This was made possible by the fall in the cost of transporting milk; the cost per litre–kilometre in 1972 was one thirtieth of that in the 1860s. In addition the invention of Lawrence's cooler, improved churns, and the use of chemical preservatives increased the time milk could be kept fresh.

There were other factors. The growth in demand for milk outran the capacity of the urban dairymen and those on the fringe of the city, and expansion to more distant regions was inevitable. In addition a series of Acts from 1853 required improvements in hygiene and in the conditions under which cows were kept in the city, and the costs of doing this made urban dairying uneconomic, and the numbers kept radically declined after the 1880s. The railways began to provide special trains to carry milk from 1887, and the rise of a number of wholesalers greatly improved the distribution of milk in the metropolis. But London's dependence on local milk was basically undercut by the rise of the railway. The same phenomenon was found in other countries. In the 1840s the outer limit of Boston's milkshed was 48 km, in 1870 104 km and by 1900 440 km. In the 1930s Berlin drew some of its milk from as far away as 700 km.[15]

Zonal shifts in the United States

At the beginning of the nineteenth century settlement and farming in the United States was still largely confined to the eastern seaboard. The following century saw a remarkable increase in population and the westward expansion of settlement. But the urbanization of the north east and rapidly rising incomes meant that a high proportion of the demand was to be found there even at the end of the century. The United States provides a clear example of the way rising demand for food products – both from the United States and from Europe – 'shunts' von Thünen zones outwards from the market.[16]

Thus at the beginning of the century livestock were grazed on the frontier, then still east of the Appalachians, and driven to market on the hoof. In parts of Pennsylvania farmers were found west of the grazing zone; but they converted their grain into whisky.[17] As agricultural settlement moved west, so livestock ranching was progressively pushed westwards, for it required cheap unenclosed range land. By the end of the century the open range had reached the drier Great Plains and the Rocky Mountains, having been displaced by more intensive farming systems. The westward movement of wheat similarly illustrates the movement of a less intensive system. On the eve of the Civil War one third of the output of wheat came from states on the Atlantic seaboard (Table 18). The east central states, between Chicago and the Appalachians, were by then the dominant producers, and so remained until the 1870s, but then declined in importance as the states west of Chicago assumed the dominant role. This region retained its

Table 18 *United States wheat output by regions, 1859–1962 (% of national output)*

	1859	1869	1879	1889	1899	1909	1921	1962
North Atlantic	14.2	12.2	7.4	6.8	5.1	4.4	4.3	1.8
South Atlantic	16.6	7.8	6.2	5.9	4.8	3.9	3.5	1.5
East north central	46.1	44.3	44.5	31.4	20.5	17.7	14.5	14.8
West north central	8.8	23.4	27.1	37.4	46.6	56.2	44.9	45.4
South central	9.9	5.0	5.3	5.2	9.4	4.7	10.1	11.5
Western	4.4	7.3	9.4	13.5	13.7	13.0	22.6	24.8

Sources: L. B. Schmidt, 'The westward movement of wheat', in L. B. Schmidt and E. D. Ross (eds.), *Readings in the Economic History of American Agriculture* (New York, 1925), p. 375; United States Department of Agriculture, *Yearbook 1921* (Washington DC, 1922), p. 523; United States Department of Commerce, *Statistical Abstract of the United States, 1964* (Washington DC, 1964), p. 655.

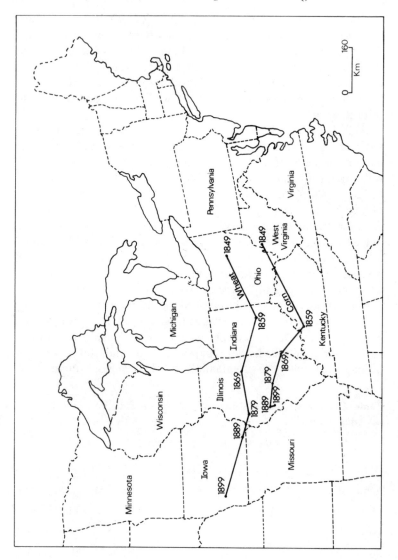

Figure 24 *The centres of wheat and maize production in the United States, 1845–99*
Sources: L. B. Schmidt, 'The westward movement of wheat', in L. B. Schmidt and E. D. Ross (eds.), *Readings in the Economic History of American Agriculture* (New York, 1925); L. B. Schmidt, 'The rise of the Corn Kingdom', in Schmidt and Ross, *Readings*, pp. 381–9.

dominance until the 1960s; but the second centre of importance, in the western states, has emerged since the 1870s. In 1859 three quarters of American wheat was produced east of Chicago, by 1962 four fifths came from west of the city. In the nineteenth century the physical centre of wheat production moved from 75 km east of Columbus, Ohio in 1849 to 113 km west of Des Moines, Iowa in 1900, a difference of 1094 km (Figure 24).[18] Wheat was displaced westwards by a variety of more intensive farming systems, one of which was the growth of maize for feeding to cattle and pigs. The corn–hog economy emerged in the 1830s in Tennessee, Kentucky and eastern Ohio, which in 1839 produced nearly one third of American maize. But the system moved westwards as settlements expanded, and the relative importance of the region of origin declined. Illinois and Indiana had become the leading producers by the 1880s, a position they retained until the 1960s when Wisconsin, Iowa and Minnesota produced nearly one third of American maize output, Indiana and Illinois less than one quarter. Between 1849 and 1900 the physical centre of production moved 772 km from near Columbus, Ohio to Springfield, Illinois (Figure 24).[19]

As settlement moved westwards milk production also moved westwards. But whereas with wheat and maize there was an absolute as well as a relative decline in the eastern states as the centre of production moved westwards, in the great zone of milk production which reached from Boston in the east to Minnesota there was instead a change in the purpose of production. In 1869 the eastern states produced not only a high proportion of all US milk, but most of its butter and cheese. By 1919 milk production in the east was five times what it had been in 1869, but the milk went to the cities for fresh consumption, and little butter or cheese was made; the western states, which lacked a nearby market for fresh milk consumption, not only delivered most of their milk to cheese and butter factories, but accounted for a high proportion of national output.[20]

Conclusions

The influence of falling transport costs upon the location of agricultural production transformed world agriculture in the late nineteenth century. On a world scale Western Europe and the eastern United States became a market for the world, and a wide zonation of production around these centres emerged. Western Europe turned to intensive products, and more extensive goods – such as wheat, wool and meat – were displaced to remote areas overseas. But the marked fall in costs and the introduction of refrigeration emancipated both milk and vegetable production from market location, and allowed areas with particularly suitable climate to specialize in such goods. The emergence of the Mediterranean as an area of specialized horticulture in the late nineteenth century illustrates this.[21]

Historians have emphasized the significance of high transport costs in pre-industrial agriculture, and the great importance of access to water routes in allowing the emergence of specialized farming – the development of viticulture exemplifies this well.[22] Somewhat less attention has been paid to von Thünen's analysis; it may be profitable to apply his ideas to certain aspects of farming before the industrial revolution. The relationship between distance from the market and the type of field system, to which he devoted so much work, might be profitably pursued.

Part Four
The pace of change

11 The diffusion of agricultural innovations

Farming has been the basis of economic life for 10,000 years or more: during this time there have been long periods when there has been little change in the crops grown, the tools used or in the institutions that govern the nature of agrarian society. But clearly there have been changes. Fourteenth-century farming in England was not the same as nineteenth-century farming, nor is agriculture in present day Russia the same as agriculture in 1900. But even in periods of undoubted change some historians have emphasized the slowness of change and preferred to write of evolution. Others have argued that there has been abrupt change, and speak of agricultural revolutions. Even among the latter there is little agreement either upon the timing and nature of 'revolution'. Thus the English agricultural revolution has been much debated. To some it took place between 1580 and 1767, with no changes thereafter; to others it only began in 1760.

There are many reasons for such disagreement. One is a failure, until recently, to allow that there is a great lag between the first appearance of a technological innovation and its general adoption by farmers; thus changes in output and productivity may be long drawn out. Another reason is a failure to define precisely what is meant by an agricultural revolution, to distinguish between causes and results, and to try and measure the rate of change. In this chapter the diffusion of innovations is considered; in the following chapter the definition and measurement of agricultural revolutions is discussed. In Chapter 13 these approaches are applied to the medieval agricultural revolution in Europe and the agricultural revolution in England.

The idea of diffusion

The idea that innovations spread slowly from a source area to other parts of the world, and that diffusion can thus explain

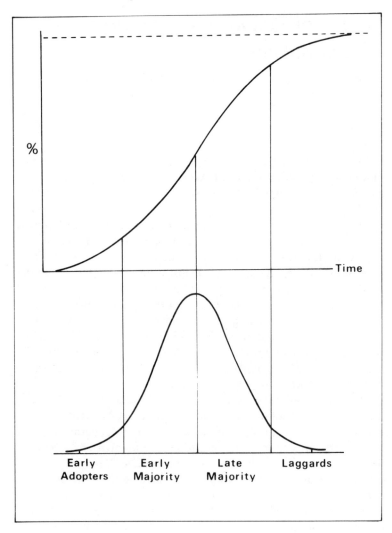

Figure 25   *The rate of adoption and types of adopters*

geographical variations in technology at any given time, was developed by anthropologists in the mid nineteenth century. In the twentieth century rural sociologists observed that farmers varied in their willingness to adopt an innovation and that in any one area there might be a considerable time lag between the first farmer's adoption of a new method, and the moment when all farmers have adopted it. Geographers have made contributions to the study of spatial diffusion, and economists, in contrast to sociologists, who have argued that the personal characteristics of farmers determine the rate of adoption, believe that expected profitability is the main cause of innovation adoption.[1] Two types of diffusion may be identified. In *social diffusion* the rate of adoption of an innovation among a particular group of farmers is studied. In *spatial diffusion* it is the geographical spread of an innovation that is the concern.

Numerous empirical studies have shown that there is a lag between the date when an innovation is first adopted, and the date when all farmers have adopted it. The rate of adoption follows an S-shaped or logistic curve (Figure 25), with initially a slow rate of adoption, then rapid, and finally a slowing rate again. British farming offers some examples of the lag. The first threshing machine was used in 1786; but the flail was not superseded until over 100 years later. Similarly the first milking machine was exhibited in 1895, yet in 1939 only 8 per cent of dairy farmers in Great Britain used the machine; but by the 1960s few were without it.[2]

Early adopters

Most farmers can be characterized by their propensity to adopt an innovation (Figure 25). A small proportion are early adopters, and an equally small proportion, 16 per cent or so, are laggards, always the last to adopt. The majority can be divided into early majority adopters and late majority adopters. Rural sociologists have made a number of generalizations about the early adopters.

First, early adopters tend to have larger farms than late adopters, are often specialized producers and have easier access to credit: the failure of an innovation has less serious consequences than for a small farmer. Second, they tend to have a high social status in their community and are well educated. Third, they often have more urban experience than laggards, and are thus used to an environment where change is typical. Fourth, early adopters are generally young – under 45 – while laggards tend to be older.[3]

Characteristics of the innovation

The social and economic characteristics of farmers may influence the rate of adoption; so too may the characteristics of the innovation itself. Thus the cost of an innovation is clearly relevant: new seed may be adopted rapidly, new buildings less so. Further, if the farmer has an implement with some years of working life left, he will be reluctant to innovate, however superior the new machine. Third, the more complex the innovation the less likely it is to be adopted. Fourth, an innovation will be adopted earlier if it is divisible: new seed or fertilizer can be tried initially in small amounts, new milking parlours cannot. Fifth, an innovation that fits easily into the existing system will be adopted before one that requires reorganization. Sixth, an innovation that saves time or reduces hard work is likely to be adopted rapidly. Finally, innovations that give prestige may be adopted rapidly.[4]

The learning process and the diffusion of information

Sociologists have argued that there are important differences in the way in which innovators and laggards learn about innovations. The innovators hear about new methods from farming journals, radio or agricultural extension workers. Laggards and late adopters hear of innovations only from neighbouring farmers. This two step flow of information has significance. The early adopters are not dependent upon their neighbours, and therefore can hear about practices that are in use some distance away. Laggards and late adopters only adopt innovations that they see their neighbours using.[5]

The process of spatial diffusion

Innovations spread spatially as well as over time among a particular group of farmers. Two types of spatial diffusion have been identified. First is *migration diffusion* where farmers who move from one area to another take with them crops, livestock and implements. Thus English emigrants took seed, animals and implements to New England and Virginia in the seventeenth century, and to Australia in the late eighteenth century. A second type of diffusion is called *expansion* or *stimulus* diffusion. Here an innovation spreads through a settled population rather like a contagious disease; it is assumed that farmers hear of an innovation only from their neighbours so than an innovation passes from its source area outwards rather like the ripples that move outwards when a stone is dropped in a pond. Within a short time of the introduction of a new farming method farmers near the source area will have adopted the innovation, but with increasing distance from the source fewer will have heard of the innovation and few adopted it.[6]

Diffusion theory and agricultural history

Until recently there were few studies of the diffusion of innovations in the past, not least because of the lack of information. Much modern study of diffusion is based upon questionnaires submitted to farmers, a method not available to the historian. Even where agricultural statistics are available they do not provide all the information necessary. Thus the agricultural census of England and Wales has recorded the number of milking machines in use in successive years; it also gives the total number of farmers in a year. But these are not all potential adopters, for not all farmers have dairy cows.

However students of diffusion have shown great ingenuity in overcoming these problems, and have demonstrated the great importance of the lag in the rate of adoption. Agricultural historians have tended to take the first recorded instance of an innovation as being significant; yet there was often a very marked lag between introduction and general adoption. Thus the turnip has often been used as an indicator of the rate of agricultural change in England but accounts of its spread have been based upon literary descriptions. However a study of probate inventories in Norfolk and Suffolk throws light upon this problem. The sample contains inventories from 1580 to 1735, and the data they contain provide information on whether turnips were grown on the farm at the time of the farmer's death, although not the area. Before 1660 very few inventories contained references to turnips; thereafter the rate of adoption was rapid until in 1710 half of the farms grew turnips. If the rate of adoption had followed the expected rate of adoption, nearly all farms would have been growing turnips by the 1760s, about a century after the beginnings of diffusion. It does not follow of course that other parts of England kept to the same sequence. The first recorded instance of turnips on farms in the Yorkshire Wolds was not until the 1750s. Thus the study of the rate of adoption of the turnip in England must take account of not only the lag over time in one place, but also spatial diffusion, the fact that the crop could not be grown in some regions (see p. 61) and that not all farmers needed to grow them.[7]

Innovators and laggards in history

Although few studies of agrarian change have been couched in terms of diffusion theory, much has been written about the characteristics of innovators in English agricultural history; but the emphasis has been not upon farmers but upon landlords. It has been argued, for example, that the sixteenth century saw the emergence of a new class of landlords imbued with the capitalist spirit, who replaced the feudal landlords who had little interest in maximizing profit. In the late seventeenth century land is thought to have been bought by successful urban merchants who invested in their new estates. According to Lord Ernle the agricultural revolution of the eighteenth century was promoted by a group of landowners – Walpole and Coke amongst them – who imposed good farming upon their tenants by means of leases and covenants. This heroic view of the age has been criticized, and more credit has since been given to tenant farmers, but there has been little systematic study of the characteristics of progressive farmers.[8] Nineteenth-century writers believed that large farmers were more likely to improve their farms than small farmers, but were rarely explicit why this should be. Recently there have been studies of farm size and the rate of adoption. In Sweden winnowing machines were adopted first on large farms when they were introduced from China in the 1750s; studies of nineteenth-century farm inventories in the same country also show that size of farm was critical in the rate of adoption. In Oxfordshire between 1800 and 1880 it has been shown that the early adopters of farm implements had above average farm size. The problems of economies of scale could however be circumnavigated. In the American midwest in the 1850s and 1860s farmers often shared the cost of buying a reaper while in the later nineteenth century the co-operative movement in the Netherlands, Denmark and Sweden helped overcome the disadvantages of small farms.[9]

Status as well as size has been regarded as a critical factor in determining the rate of adoption; thus freeholders are thought to have adopted innovations more readily than tenants in the seventeenth-century Netherlands. In England freeholders were few by the eighteenth century, and contemporaries and historians have put more emphasis upon the possession of long leases and tenant right. The former gave the tenant the time to reap the reward for investment; the latter compensated the outgoing tenant for improvements and thus gave a similar security. There have however been few studies testing this difference.[10]

In modern farming it is the farmer that makes the decision to innovate; in traditional agriculture this was not entirely so. Even good tenants were often bound by covenants and leases; but wherever the open field system prevailed there were collective regulations, which the individual had difficulty in avoiding, that governed the rotation of crops, while the persistence of common grazing made selective breeding impossible. Thus the laggard controlled the overall rate of change, and it is for this reason that historians have put so much emphasis upon the extinction of the open field system as a prerequisite for agricultural progress.

The learning process and the diffusion of information

Sociologists have argued that innovators learn about innovations from the radio and from books, laggards and late adopters only from their neighbours. How did innovators learn about new methods in the past?

Books can have had only a limited role in the diffusion of information until very late on. It was not until 1523 that a book on agriculture was printed in England: twenty-two were published in 1600–50, fifty in 1650–1700, and the same number in 1700–50. But between 1750 and 1790 over 100 were published, and between 1750 and 1850 the number increased fivefold and the number of farming journals rose tenfold. In France there was also an increase after 1750.[11] But not all books on agriculture were useful, and not all farmers could have read them. In the East Riding in the 1750s only half the population could sign the parish register. In sixteenth-century France only 10 per cent of the labourers in the Narbonne could sign their name, and only one third of even the wealthier farmers. As late as 1871 40 per cent of the Catholic population of Ireland were illiterate and in 1881 65 per cent of the rural population of Italy could neither read nor write. As to the proportion of *farmers* who could read or write at different dates, it is impossible to say.[12]

Other means of spreading knowledge were established in the late eighteenth century. The first agricultural society in England, the Bath and West, was founded in 1778, but the national society, the Royal Agricultural Society of England, not until 1838. No doubt the shows and journals promoted by these organizations spread information, but probably more important were local farmers' associations: there was one such association in 1750, thirty-five in 1800 but nearly 600 by the 1870s. But it is likely that until the twentieth century the most important means of spreading

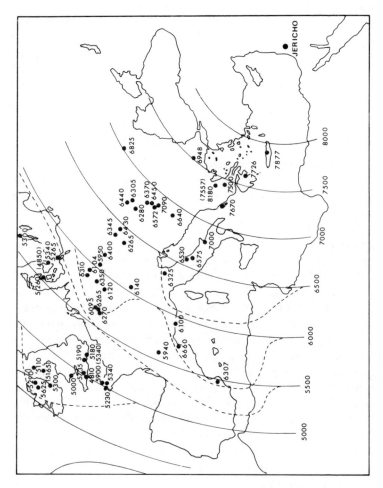

Figure 26   *The diffusion of wheat and barley into Europe*
         Note: Dates are in years before the present. Arcs show the
         expected position at 500 year intervals.
         Source: A. J. Ammerman and L. L. Cavalli-Sforza, 'Measuring
         the rate of spread of early farming in Europe', *Man*, vol. 6
         (1971), pp. 674–88.

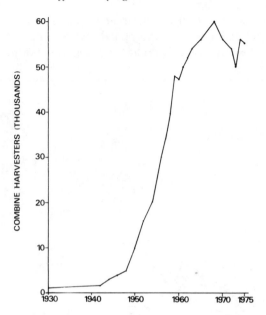

Figure 27 *Growth in the number of combine harvesters in England and Wales, 1930–75*
Source: Ministry of Agriculture, *A Century of Agricultural Statistics: Great Britain 1866–1966* (London, 1968), p. 73.

information about new methods was by observing neighbours' practices and discussion with other farmers. By the late eighteenth century farmers in England were visiting other districts to learn, and hiring labourers with specific skills to teach their own workers. In 1825 Flemish reapers demonstrated the Hainaut scythe in many parts of Scotland, although with little success.[13]

Spatial diffusion

In the developed countries the importance of radio, the printed word and the agricultural extension officer has greatly increased, so that an increasing proportion of all farmers hear of the innovation directly rather than from their neighbours. This means that innovations no longer spread slowly through the countryside from neighbour to neighbour;[14] but in the past such spatial diffusion was more common, and there are many instances of such a migration of ideas and methods. The very domestication of plants and animals spread slowly from the Near East through Europe. The introduction and spread of the plough has also been traced; the three field system is thought to have its origins in northern France

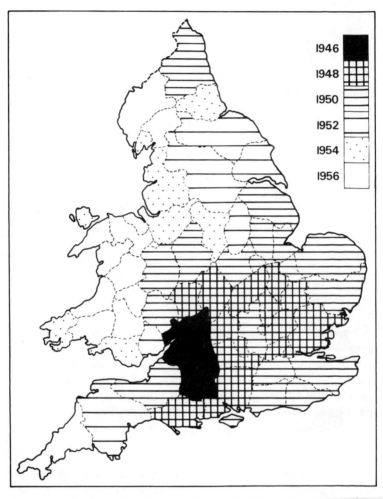

Figure 28 *The diffusion of the combine harvester in England and Wales to 1956*
Note: In 1956 each county had reached one combine per 400 ha of cereals
Source: G. E. Jones, 'The diffusion of agricultural innovations', *Journal of Agricultural Economics*, vol. 15 (1963), pp. 387–405.

in the ninth century and diffused from there to the rest of Europe. There have been attempts to trace the spread of the potato within Europe, of the diffusion of tobacco, and of the adoption of hybrid maize in the United States. Unfortunately all such efforts lack the information that is essential to map and model the diffusion of agricultural innovations.[15] Two examples however may illustrate this approach.

The recent advances in radiocarbon dating yield dates of the first evidence of the Neolithic at fifty-three sites in Europe.[16] Assuming that the early domesticates of wheat and barley were carried by migratory peoples rather than by stimulus diffusion, it is possible to simulate the spread of agriculture from the Near East into Europe, beginning about 8000 BC and reaching the north west in c. 5200 BC. Plant domestication spread at about 1 km per year, and apparently spread at an even rate along the innovation front rather than following the Danube into the heart of Europe (Figure 26).

A much more recent example, and thus better documented, is the adoption of the combine harvester in England and Wales.[17] The first combines were introduced in 1926–8. There was a very slow increase from 1930 to 1948, the reaper and the thresher still being the dominant method of harvesting. But from 1948 to the 1960s there was a very rapid increase, which slowed down after the 1960s, thus approximating an S-shaped curve (Figure 27). The early spread of the machine is shown in Figure 28. The first combine was used in Gloucestershire in 1928, but twenty years later only Gloucestershire and Wiltshire had at least one combine per 400 ha in cereals. The combine spread outwards from this source region, but not at an even rate. West of Gloucestershire farms were small, mainly livestock producers and often on sloping land; eastwards cereal growing was the dominant form of enterprise, fields and farms large, and the terrain even, so that East Anglia adopted the machine more rapidly than its position in relation to Gloucestershire might suggest.

Conclusion

Recent work on diffusion theory in agricultural history suggests it may make an important contribution to the understanding of agrarian change. This theme is returned to in Chapter 13.

12 The definition and measurement of production and productivity growth in agriculture

Although most historians have emphasized the slowness of change in agriculture, there have been periods when it has greatly accelerated, and these phases have been dubbed periods of agricultural revolution. Unfortunately there is often little agreement as to when these crucial changes began. This may be partly explained by a failure to distinguish between the introduction of a new farming method and its general adoption, as has been noted (pp. 153–63). More commonly it is due to a failure to specify exactly what is meant by an agricultural revolution, and how the rate of change is to be measured. But such problems are dealt with by economists in their studies of modern agricultural change, and it may be that their approaches can be applied to the study of agricultural change in the past.

Some semantic difficulties

Part of the problem of defining the course of an agricultural revolution arises from failure to specify what is changing and where. Much of the historiography of the agricultural revolution in Europe has been concerned with the obstacles to improvement, and the rate at which new methods have been adopted. As has been seen earlier (pp. 153–63) there may be a very considerable lag between the introduction of a new technique and majority adoption. There will thus be a lag between the innovation and the time when it influences regional or national productivity. Further, few periods of rapid agricultural change are characterized by the adoption of only one new farming method. More usually the adoption of a number of new inputs, commonly interlocked, is characteristic. They may however spread at different rates. Thus the success of the green revolution in Asia in the 1960s and 1970s was not a matter of merely adopting an improved variety of wheat or rice. To increase yields it was necessary also to have irrigation, fertilizers and pesticides. The unequal geographical spread of these inputs accounts for the very varying success of the new hybrid varieties.

It is also necessary to specify the area that is being studied when studying the course of an agricultural revolution. Thus early historians spoke of an agricultural revolution in England after 1760; yet the evidence of an accelerated adoption of new techniques was drawn mainly from eastern England, and within that region, from East Anglia. Even within this region techniques were adopted at different rates upon light soils and heavy soils (see pp. 60–5). The new techniques were essentially concerned with arable farming. Yet much of midland and western England, and in particular upland England, had little cropland, and for these areas the agricultural revolution of the eighteenth century had limited significance.

There has also been a failure to distinguish between what B. F. Johnston and J. T. Nielsen have called 'proximate' and 'conditioning' factors.[1] By 'proximate' factors they mean those changes *on the farm* that cause an increase in output or an increase in productivity; such might be the sowing of an improved crop variety that increases yields or the replacement of the sickle by the reaper so increasing labour productivity. 'Conditioning' factors are changes in the institutional and economic environment which encourage the adoption of new techniques on the farm. Thus the provision of agricultural credit facilitates investment by farmers, but does not cause it. Enclosure, to which English historians have given so much emphasis, did not cause the adoption of new farming methods, it simply made change easier. It is worth recalling that much of Western Europe experienced little or no enclosure in the nineteenth century, but did achieve a rapid growth of output.[2]

A further cause of confusion has been to assume relationships between the growth of agricultural output and other components of the economy. Such a view has been taken by a number of modern development economists, who have argued that industrialization is only possible if preceded by changes in agriculture. This is because agriculture is seen in an underdeveloped economy as the only source of capital, labour and of entrepreneurs for industry; therefore it is assumed there must be advances in agricultural productivity before industrialization can get under way. This view has been clearly stated by Paul Bairoch, who argues that the agricultural revolution in Europe was a comparatively short period when output per head of the agricultural population rose from 125 per cent of the consumption needs of the total population to 150 per cent. This reduced the risk of harvest failure, reduced death rates, and allowed a sustained increase in the industrial population. The revolution covered a period of 40–60 years and began in

Table 19 *Growth of agricultural output in the nineteenth century; selected countries*

1 *France*	*percentage per annum*	2 *United States*	*percentage per annum*
1803–12 to 1813–24	0.2	1800–1810	3.0
1815–24 to 1825–34	1.2	1810–20	2.9
1825–34 to 1835–44	1.5	1820–30	3.2
1835–44 to 1845–54	1.1	1830–40	3.5
1845–54 to 1855–64	1.4	1840–50	2.3
1855–64 to 1865–74	0.8	1850–60	3.6
1865–74 to 1875–84	−0.05	1860–70	1.9
1875–84 to 1885–94	0.35	1870–80	4.3
1885–94 to 1895–1904	0.9	1880–90	1.8
		1890–1900	2.4

3 *Great Britain*	*percentage per annum*	4 *Great Britain*		*percentage per annum*
1801–11 to 1831–41	1.2	1867–69	to 1870–76	0.7
1811–21 to 1841–51	1.5	1870–76	to 1877–85	−0.5
1821–31 to 1851–61	1.8	1877–85	to 1886–93	0.5
1831–41 to 1861–71	1.3	1886–93	to 1894–1903	0.1
1841–51 to 1871–81	0.7	1894–1903	to 1904–10	0.7
1851–61 to 1881–91	0.5	1904–10	to 1911–13	0.2
1861–71 to 1891–1901	0.7	1911–13	to 1920–22	−0.5
		1920–22	to 1933–39	−3.1
		1923–29	to 1930–34	0.6

Sources:

1 Net agricultural output: J. C. Toutain, *Le produit de l'agriculture français*, (Paris, 1961), vol. 2 pp. 127–9.
2 Real farm product: M. W. Towne and W. D. Rasmussen, 'Farm gross product and gross investment in the nineteenth century', in *Trends in the American Economy in the Nineteenth Century*, Studies in Income and Wealth, vol. 24 (Princeton, NJ, 1960), p. 260.
3 Real product: P. Deane and W. A. Cole, *British Economic Growth, 1688–1959: trends and structure* (Cambridge, 1962), p. 170.
4 T. W. Fletcher, 'Drescher's index: a comment', *Manchester School of Economic and Social Studies*, vol. 23 (1955), p. 181.

England in 1690–1700, in France 1750–60, in Switzerland 1780–90, in Germany and Denmark 1790–1800 and in Austria, Sweden and Italy in the 1820s. In each country the agricultural revolution preceded the industrial revolution. Whatever the truth of the relationship between agrarian change and industrialization, it is apparent that such an assumption presumes an agricultural revolution, and if presumed, it may be found.[3]

Perhaps the most powerful criticism that may be made of the present study of agricultural revolutions is that too little attention is paid to the *results* of technological change. Much of the historiography of agrarian change deals with the adoption of new farming methods, and the reasons for the slowness of their adoption. Yet properly to measure the pace of agrarian change attention should be focused on the *outcome* of technological change. Did it increase yields, did output rise and was productivity increased?

It is possible to frame various definitions of an agricultural revolution in terms of these concepts. First, it could be defined as a period when there is an increase in the rate of growth of the volume of agricultural production. Second, it could be defined as a period when the rate of increase of the volume of agricultural output exceeds the rate of increase of total population. Many historians – like Bairoch – have seen the eighteenth and nineteenth centuries in these terms. For the first time food output ran ahead of population increase and ended the Malthusian trap.[4]

But an increase in the volume of output does not require any increase in productivity; it can be achieved by the expansion of the cultivated area and by increased labour inputs; yet most discussions of agricultural revolutions assume that there are productivity increases, either of output per head or output per hectare or of both. This gives a third possible definition, that an agricultural revolution is a period when the rate of growth of productivity rises. We turn now to consider how the volume of output may be measured, then to concepts of agricultural productivity.

The measurement of agricultural output

The measurement of agricultural output is difficult for several reasons. First, agricultural censuses in Europe date only from the mid nineteenth century at the earliest, and they frequently record only the area in crops and the number of livestock kept. If yields are known, then an approximation of arable output can be made but livestock production cannot be accurately derived simply from the numbers of stock. Second, agricultural output is highly heterogeneous: it may include wool, olive oil, wine, chestnuts, bacon and woad. How is this diversity to be reduced to one unit of account? Third, the compilation of national output figures must exclude double counting. Thus grain is a major output of modern British farming, but much of it is fed on farms to livestock, and appears in the form of meat, eggs, milk and wool.

In the event several ways of measuring output have been devised. First, crop and livestock output can be reduced to a common physical unit, such as starch equivalents, wheat equivalents or direct calories. The only historical series to use this method is that by Paul Bairoch, who estimated output in direct calories for a number of European countries at various dates between 1810 and 1910, but unfortunately published only output per agricultural worker. The principal objection to this method is that it does not include non-food products.[5] A second method is to measure the volume of output of the different products, but weight this by price. Such a method is used by FAO to measure trends in food and agricultural output, and also by many government offices to measure national output. Two difficulties arise. First, weighting must be in constant prices otherwise inflation or deflation will give spurious changes in output. Second, international comparisons require that prices must be expressed in a common currency. This may be misleading. A recent study of relative agricultural productivity within Western Europe gives different rankings according to the national currency in which inputs and outputs are expressed.[6]

A third and radically different way of expressing trends in agricultural output is to estimate income generated in the agricultural sector. This has been done by Phyllis Deane and W. A. Cole for nineteenth-century Britain, while M. Towne and W. D. Rasmussen have measured the contribution of the American agricultural sector to the United States national income in the nineteenth century. The latter estimate includes the rental value of farm houses, home industries and land improvements, and is thus based on concepts very different to the physical measurement of agricultural production[7] (Table 19).

The only comprehensive records of the growth of agricultural output are those made by FAO (Table 20) for all major regions from 1952–4 to 1970–6 and annually from 1961 to the present.[8]

Various estimates for the nineteenth and early twentieth centuries are available for Britain, France and the United States (Table 19). For the eighteenth century there are only J. C. Toutain's estimate of the growth of French net agricultural output, and estimates of the growth of British corn output. Bearing in mind the doubtful statistical base of many of these calculations, the methodological problems of constructing indices, and the fact that some are estimates of physical output, others of income, the data should not be pressed too hard. But some points can be made.

First the growth of agricultural output in the eighteenth century was slow: 0.36 per cent per annum in England, and 0.5 per cent per annum in France;[9] the latter figure is thought too high by some French writers.[10] These rates of increase were not higher than population growth. In both countries output rose at over 1 per cent per annum from 1801 until the mid nineteenth century, and at below 1 per cent in the second half of the nineteenth century (Table 19). In the United States output grew much more rapidly, the lowest rate in a decade being 1.8 per cent per annum, the highest 4.3 per cent per annum. Between 1880 and 1950 output grew at 1.8 per cent per annum in the United States, 1.6 per cent in Denmark, 1.1 per cent in France, and 0.8 per cent in Britain.[11] But in the 1950s and 1960s agricultural output in Western Europe grew at between 2 and 3 per cent per annum, and even higher rates were

Table 20 *Rates of growth of total agricultural output by major world regions, 1952–4 to 1970–6 (% per annum)*

	1952–4 to 1959–61	1959–61 to 1967–9	1952–4 to 1967–9	1970–6
Western Europe	2.4	2.7	2.6	1.6
North America	1.4	1.8	1.6	2.8
Eastern Europe & USSR	5.1	3.2	4.1	2.0
Oceania	3.3	3.2	3.3	1.3
Developed countries	2.9	2.6	2.7	2.1
Latin America	3.4	2.4	2.9	2.9
Far East	3.3	2.5	2.9	2.6
Near East	3.4	3.1	3.2	3.9
Africa	2.6	2.4	2.5	1.1
Developing countries	3.2	2.5	2.9	2.6

Sources: FAO, *World Agriculture, the Last Quarter Century*, World Food Problems, no. 13 (Rome, 1970), p. 9; FAO, *The Fourth World Food Survey* (Rome, 1977).

obtained in other developed regions, while rates were also very high in developing countries (Table 20). Thus it can be seen that the growth of output was low in the eighteenth century, higher in the first half of the nineteenth century, low again – but not to eighteenth-century levels – in the second half of the nineteenth century. Output grew more rapidly in the first half of the twentieth century, and then reached unprecedented rates in Western Europe in the postwar decades.

It should be noted that much of the high rate of increase in output in the nineteenth-century United States was attributable simply to the rapid expansion of the cultivated area, for yields did not increase. It is also likely that European output was expanded primarily by increasing the crop area in the twelfth and sixteenth centuries. It is questionable as to whether mere increase in the rate at which the volume of output increases is a sufficient definition of an agricultural revolution. It is also noteworthy that the eighteenth-century increases in France and England could have been little in advance of population increase. It is necessary now to consider productivity changes.

Concepts of productivity

Agricultural productivity is a term widely used but not easily defined; nonetheless most economists would agree that the term 'agricultural revolution' assumes more than the acceleration of the volume of output; productivity must also rise. Productivity has been defined as 'the increased efficiency with which inputs are processed into outputs'.[12] The ideal measure is *total factor productivity* which relates total input to total output. R. E. Gallman has defined total factor productivity growth as the rate of growth of output minus the rate of growth of aggregate factor input, each factor growth rate being weighted by the share of the factor in total agricultural income. Gallman concluded that total factor productivity rose by 0.56 per cent per annum in the nineteenth century in the United States, which may be compared with a rate of 1.7 per cent per annum for Britain between 1949–50 and 1959–60. However the United States example is the only published historical example of total factor productivity growth.[13]

Far more commonly used are partial productivity measures, where output changes are related to changes in one factor of production, most commonly to land and labour.

Labour productivity

Labour productivity is the most used measure of productivity growth, and relates total output to the total labour force, or more commonly, the total male labour force. Unfortunately, such data are rare in the past, and changes have often been measured by describing the labour saving resulting from the introduction of a particular implement (see pp. 131–4). In the eighteenth century the output of corn per caput of the male labour force rose at 0.2 per cent per annum in Britain, more rapidly in the first half than the second half, while in France the net agricultural output per head rose at 0.3 per cent per annum, although some would set this much lower, at between 0.1 and 0.2 per cent.[14] In the nineteenth century labour productivity in Britain rose at 1 per cent per annum, a result of a decline in the labour force after 1851, and of increasing output 1800–50, and in France at between 0.25 and 0.5 per cent per annum.[15] Rather higher rates were achieved in the early twentieth century (Table 21) typically between 1 and 1.5 per cent per annum and the rate of increase steadily rose until it reached increases per annum of the order of 4–6 per cent in Western Europe after 1950 (Table 22). By then not only was the volume of output increasing rapidly, but the agricultural labour force was everywhere in decline. It should be noted that even in the developing world, where labour forces were increasing, labour productivity was rising at rates above those achieved in nineteenth-

Table 21 *Rate of increase in output per male worker in agriculture, 1880–1950 (% per annum)*

	United States	Denmark	France	United Kingdom
1880–90	0	1.7	1.3	0.8
1890–1900	1.8	1.4	−0.4	0.25
1900–10	0.5	1.8	1.5	−0.2
1910–20	1.0	−0.5	1.0	0.7
1920–30	1.75	4.1	2.6	0.7
1930–40	2.5	−0.4	–	2.2
1940–50	4.9	3.2	0.8*	2.4

*1930 to 1950

Source: Y. Hayami and V. W. Ruttan. *Agricultural development: an International Perspective* (London. 1971), pp. 327–31.

Table 22 *Rate of increase of agricultural production per agricultural worker (% per annum)*

	1961–5 to 1970	1970–6
North America	5.4	6.6
Western Europe	6.3	4.7
Oceania	3.5	2.4
Eastern Europe & USSR	6.2	4.3
Other developed countries	5.7	4.9
All developed countries	5.7	4.8
Africa	1.4	−0.2
Latin America	2.2	2.0
Near East	2.2	2.9
Far East	2.1	1.4
Asian centrally planned economies	2.3	2.1
Other developing countries	0.9	0.0
All developing countries	2.1	1.7
World	2.2	1.8

Source: FAO, *The Fourth World Food Survey* (Rome, 1977), p. 14.

century Europe. Thus labour productivity, unlike the volume of output, has experienced a continuous and uninterrupted increase since the eighteenth century.

Land productivity

The productivity growth of land is measured by relating the changes in total agricultural output to changes in the area used for agriculture. Quite apart from the problems of measuring total output (see pp. 167–9) there are methodological problems in defining the agricultural area. Is it, for example, to include arable, pasture and rough grazing? Are allowances to be made for multiple cropping? Is bush fallow to be included in estimates of the agricultural area in Africa and in other countries where shifting agriculture is common? Not surprisingly such estimates are rare. V. W. Ruttan and Y. Hayami have made estimates for four western countries from 1880 to 1960; in all there was a slow rate of increase in land productivity until after 1940 when rates of increase reached to between 2 and 4 per cent per annum (Table 23).

In the absence of such estimates, historians have resorted to the use of the yields of individual crops to show long term trends in land productivity. This has drawbacks. It neglects the increase in livestock output which was undoubtedly a major feature of nineteenth-century Europe; further, data are commonly only available for wheat, which may be a small proportion of the total crop area. Nor are there any reliable series of wheat yields before the middle of the nineteenth century. The data however confirm the rapid increase in the last thirty years (Figure 20), and the differences between Western Europe and the overseas producers which were noted earlier (see pp. 130–1). Prior to 1850 information on crop yields is sporadic and speculative; some of the evidence for England is discussed later (see pp. 184–6). Although modern crop yields are normally expressed in terms of the amount harvested per hectare, yields in earlier periods can often only be obtained from relating the amount sown to the amount harvested. Slicher van Bath has summarized changes in these seed/yield ratios from the early sixteenth to the early nineteenth centuries for the major regions of Europe (Table 24). His data have not been without criticism. They show however that the differences in yield between the major regions which existed in the sixteenth century persisted – indeed widened – into the early nineteenth century, that there was little increase in southern, central and eastern Europe between 1500 and 1820, and even in Britain and the Low Countries yield increases in the eighteenth century were comparatively modest. The differences between regions established in the sixteenth

Table 23 *Rate of increase in agricultural output per hectare (% per annum)*

	United States	Denmark	France	United Kingdom
1880–90	0.3	1.1	0.4	0.2
1890–1900	−0.5	1.3	0.4	−0.5
1900–10	0.2	2.9	0.6	0.35
1910–20	0.0	−0.2	0.4	0.2
1920–30	0.9	4.0	1.4	0.5
1930–40	0.4	−0.6	–	1.7
1940–50	1.2	2.3	0.2*	1.4
1950–60	2.1	2.9	4.9	1.9

*1930 to 1950.

Source: Y. Hayami and V. W. Ruttan, *Agricultural Development; an International Perspective* (London, 1971), pp. 327–31.

century persisted into the later nineteenth century and indeed until the present day; as can be seen the rankings of countries have not radically changed over the last century (Table 25). The highest yields, from the sixteenth century to the present day, have been in north western Europe, the lowest in the south and the east. Whether this can be related to a decline in intensity away from the major urban centres in Europe, or to differences in precipitation and soil type, is a matter for debate.

The sources of productivity growth

It is always possible to increase agricultural production by increasing labour, capital and land inputs. But this does not necessarily lead to an increase in productivity. Until recently most discussions of agricultural change emphasized the importance of technological change in causing increases in output and in productivity. However, attempts by American economists to measure the contribution of new inputs to output increases has suggested that there must be other causes of productivity growth, for increases in output per unit of input cannot be explained simply by the growth of new inputs.[16] No entirely satisfactory explanation of this residual has been put forward, but some

Table 24 *Seed/yield ratios* in the regions of Europe, 1500–1820*

	1	2	3	4
1500–49	7.4	6.7	4.0	3.9
1550–99	7.3	–	4.4	4.3
1600–49	6.7	–	4.5	4.0
1650–99	9.3	6.2	4.1	3,8
1700–49	–	6.3	4.1	3.5
1750–99	10.1	7.0	5.1	4.7
1800–20	11.1	6.2	5.4	–

Zone 1: England, the Low Countries
Zone 2: France, Spain, Italy
Zone 3: Germany, Switzerland, Scandinavia
Zone 4: Russia, Poland, Czechoslovakia, Hungary

*For wheat, barley, oats and rye.

Source: B. H. Slicher van Bath, 'Agriculture in the vital revolution', in E. E. Rich and C. H. Wilson (eds.), *The Cambridge Economic History of Europe*: vol. 5, *The Economic Organization of Early Modern Europe* (Cambridge, 1977), p. 81.

Table 25   *Wheat yields in Europe, 1850–1971 (100 kg/ha)*

	1850	*1909–13*	*1948–52*	*1969–71*
Denmark	c.12(1)	33.1(1)	36.5(1)	45.76(2)
Belgium	10.5(2)	25.3(2)	32.2(3)	40.58(5)
Netherlands	10.5(3)	23.5(4)	36.5(2)	46.2(1)
Germany	9.9(4)	24.1(3)	26.2(5)	41.5(4)
United Kingdom	9.9(5)	21.2(5)	27.2(4)	42.2(3)
Austria	7.7(6)	13.7(6)	17.1(8)	32.73(7)
France	7.0(7)	13.1(7)	18.3(7)	36.5(6)
Italy	6.7(8)	10.5(11)	15.2(9)	23.9(11)
Norway	5.7(9)	16.6(8)	20.6(6)	31.3(8)
Romania	–	12.9(10)	10.2(12)	17.5(13)
Hungary	–	13.2(9)	13.8(10)	26.5(10)
Bulgaria	–	6.2(15)	12.4(11)	28.3(9)
Spain	4.6	9.2(13)	8.7(14)	12.7(15)
Greece	4.6	9.8(12)	10.2(13)	18.5(12)
Russia	4.5	6.9(14)	8.4(15)	14.2(14)

Sources: J. Blum, *Lord and Peasant in Russia from the Ninth to the Nineteenth Century* (Princeton NJ, 1961), p. 330; League of Nations, *International Statistical Yearbook 1926* (Geneva, 1927), pp. 38–9; FAO, *Production Yearbook, 1962,* vol. 16 (Rome, 1963), pp. 34–5; *1977,* vol. 31 (Rome, 1978), pp. 94–5.

possible reasons can be noted. First, modern agriculture has seen changes in the quality of physical inputs; thus simply measuring increases in the amount of fertilizer applied to the land neglects the fact that the plant nutrient content of a given tonnage has risen.[17] Second, although the labour input in agriculture has declined, the quality has improved; the modern farm labourer is better educated and more capable of dealing with machinery than his predecessor. Third, the quality of farm management has improved, partly due to the better education of farmers, but also to the growth of farm extension services, which now offer advice upon not only scientific and technical advances but on farm accountancy and other management problems. Fourth, many economists have emphasized the importance of 'learning by doing'. The adoption of a new method does not give a once and for all increase in output and productivity; it takes time to learn the best way of using a new method and so productivity increases are spread out over a longer period.[18] Fifth, increases in labour productivity in a country may be obtained when the proportion of the area occupied by large farms increases and that of small farms declines. This is because economies of scale make possible labour productivity increases on the national farm as the area in large farms increases. It does not necessarily give increases in output per hectare. Recent work suggests that yields are greater upon small than large farms.[19] Sixth, productivity increases on the national level may be expanded by increased regional specialization, where the production of a crop is concentrated in that region in which it has a comparative advantage; some of the increase in maize yields in the United States has been due to a progressive concentration of the crop area in the mid west, which is physically most suited to its cultivation.

None of this means that technological advance is not an important – indeed the most important – factor in accounting for increased productivity. New methods can raise the production function so that output per unit of input is permanently increased.

Most modern attempts to measure the rate of growth of agricultural output and productivity depend upon statistical data which are not available before the middle of the nineteenth century; the methods of measurement also have areas of considerable methodological uncertainty. Nonetheless it is worth reappraising the agricultural revolutions of the European past in the light of these modern approaches.

13 Agricultural revolutions in Europe

Agricultural revolutions in Europe

The term agricultural revolution has been applied to various places, and at various times; and it clearly means different things to different historians. In this chapter we examine two agricultural revolutions – in medieval Europe and in the eighteenth and nineteenth centuries in England – in the light of the approaches discussed in the two preceding chapters.

The medieval agricultural revolution

In the last centuries of the first millennium AD there was only a slow increase in the population of Western Europe, but between approximately AD 1000 and 1340 there was a substantial rise in numbers. The population of England, for example, may have tripled between 1100 and 1340, from 1.5 million to 4.5 million, and that of France and Germany doubled. This was also a period of urban and economic development, and an increasing volume of trade. But the demographic and economic revival of the high Middle Ages was only possible, some historians have argued, because the period was preceded by a number of agricultural innovations whose adoption constituted an agricultural revolution. These innovations produced an increase in output which in turn allowed a greater population.

Georges Duby and Lynn White Jr. are the leading advocates of this point of view, but their revolutions cover different periods.[1] According to Duby the revolution began in the late eighth century AD, was at its most intense between 950 and 1050, and came to an end in the twelfth century, after which there were no major innovations in agriculture until the eighteenth century. White, on the other hand, sees the revolution beginning in the sixth century and ending in the ninth century. The innovations they consider crucial are shown in Table 26. Both White and Duby place greatest emphasis upon the changes in rotations, or field systems, the replacement of the ox by the horse, and the improvement of the plough.

Since Duby and White wrote, the origins of the common fields have been subject to a considerable reappraisal. Far from being developed *en bloc* by the Anglo-Saxons in the middle of the first millennium, and spreading from there throughout Europe, they do not seem to have reached maturity until the twelfth and thirteenth centuries and even then were not found throughout Europe.[2] The mature system was characterized by unfenced arable land, whose stubble and fallow were grazed in common, the division of the arable fields into narrow and scattered plots, the grazing in common of land outside the arable fields, and the collective regulation of the husbandry of the arable land. It is now argued that as late as the sixth century most West Europeans lived in hamlets and isolated dwellings, farmed their arable holdings in one unitary block, and grazed the intervening pasture in common. It was only the growth of population and the occurrence of partible inheritance that led to nucleated villages, fragmented arable, and collective regulations.

Of these regulations, White and Duby stress the shift from a two course to a three course rotation. In the two field system half the

Table 26 *Agricultural innovations in the medieval period*

	G. Duby		L. White Jr
1	Water-mill for grinding wheat	1	–
2	Windmill for grinding wheat	2	–
3	Horse-shoe and horse-collar	3	Horse-shoe and horse-collar
4	Improved yoke for oxen	4	–
5	Increased use of iron in farm implements	5	Increased use of iron
6	Diffusion of the heavy plough	6	Diffusion of the heavy plough
7	Three course rotation	7	Three field system
8	–	8	Growth of legumes
9	–	9	Improved felling axe
10	Replacement of ox by horse	10	Replacement of ox by horse

Sources: G. Duby, 'La révolution agricole médiévale', *Revue de Géographie du Lyon*, vol. 29 (1954), pp. 361–6; Lynn White Jr, *Medieval Technology and Social Change* (Oxford, 1962), pp. 40–78.

arable land was in an autumn sown crop, the other half left fallow; this was reversed the following year. This system had its origins in the Mediterranean where it was necessary to use the rainfall of two winters to get an adequate crop. The lack of spring and summer rain prevented the cultivation of spring sown crops. The two field system was found widely in northern Europe, although rainfall was sufficient for spring sown crops. The switch to the three field system had considerable advantages. It increased the amount of arable under crops in a year from one half to two thirds; it allowed the growth of oats, a spring sown crop, and an important food for horses, and permitted the cultivation of peas and beans. The latter not only helped fix nitrogen in the soil, but gave a protein rich food, and may have made men fitter and stronger. Although the first evidence of the three field system occurs in northern France in the ninth century, its diffusion seems to have been slow, and by 1250 it was only well established in parts of the Paris basin. In England the two and three field systems coexisted in the thirteenth century, but there are signs that, possibly due to population growth, there was conversion of two field to three field systems. Whatever the advantages of the three field system, it did not replace the two field system until long after the periods of agricultural revolution identified by White and Duby.[3]

Many historians have stressed the importance of improvements in the plough. The earliest ploughs, or *ards*, were designed to deal with the dry and friable soils of the Mediterranean lands, and consisted of a beam, a stilt and a share that pierced the upper soil. Fields had to be cross-ploughed to get an adequate tilth. The late medieval plough however was a more sophisticated and heavier implement. It had a coulter, a vertical metal bar that cut the turf in advance of the share, a share, and a mouldboard that turned the sod cut by the share. It required a heavy frame to hold these parts, and four or more oxen to pull it; in some cases it had wheels. The contact parts were increasingly made from iron. This plough was far more powerful than the *ard*. It made cross-ploughing unnecessary, gave a better tilth, allowed the easier cultivation of heavy clay soils, and also made possible the formation of ridge and furrow which gave a rudimentary surface drainage. There has been a spirited controversy over the origins of the heavy plough. Some believe it existed in Roman times, others that it was introduced into England by the Anglo-Saxons, although this is now discounted by most authorities. The earliest evidence of the mouldboard plough in England is in the twelfth and thirteenth centuries, and not before. This suggests that while the period of the medieval agricultural revolution saw the introduction of the heavy plough, its majority adoption came later.[4]

The *ard* and the heavy plough were drawn by oxen. The horse was not only expensive, but could not be used for ploughing until the invention of the horse-shoe and the horse-collar which allowed it to exert traction. Both had been invented by the ninth or tenth centuries. The horse had no greater traction power than the ox, but it could carry out similar tasks more rapidly; this allowed more frequent ploughings, which reduced weeds and created a better tilth. It also made easier the task of harrowing in spring, which had to be done rapidly. There is little reliable evidence on the chronology of the replacement of oxen by horses. The horse is thought by some to have replaced the ox in the Paris basin and in Flanders by the twelfth century, but was not commonly used for ploughing in England until the sixteenth century.[5] This suggests that the ox was not replaced by the horse during the medieval agricultural revolution. Indeed the ox was still of greater importance than the horse in France in the eighteenth century, except in Picardy, while the ox only gave way to the horse in the Netherlands in the nineteenth century.[6]

Peas, beans and vetch all have nodules on their roots that help fix nitrogen in the soil; they were the only leguminous crop known to the medieval farmers of Western Europe. They were also an important food crop, and could be grown on the fallow. But most references to these crops occur after 1250, and references to them increase in frequency in the fourteenth century. In some instances they occupied a significant proportion of the arable. Thus in Artois in 1314 they made up 20 per cent of the land sown in spring. In contrast on the Winchester estates in southern England they were only 0.97 per cent of the sown area in 1314, although this rose to 8.3 per cent in 1345. It is thus doubtful if these crops were of much significance until the later Middle Ages, and could have hardly played a major role in any agricultural revolution dated before the thirteenth century.[7]

There were other ways of increasing soil fertility, and the most important was to increase the supply of manure. This required better feed and more numerous animals, and some means of concentrating manure on the fallow before the sowing of autumn crops. There is some evidence that population growth in the thirteenth century led to an expansion of the cropped area into the areas of common grazing, which reduced the fodder supply and hence also the number of livestock. Sheep-folding was practised on some light upland soils in England in the thirteenth century and there is some evidence of stall-feeding but these practices were not widespread. Thus it is difficult to show that there was any substantial increase in the supply of manure in the Middle Ages, and some would argue that it declined in the thirteenth century.[8]

Was there then a medieval agricultural revolution? There is of course no evidence upon trends in the volume of output; but there is abundant evidence that the area in cultivation increased substantially between the tenth and early fourteenth centuries, and the population of Western Europe certainly increased. But it is far from clear that output per head of either the total population or the agricultural population increased in this period. Indeed many have argued that output per head was falling in the late thirteenth century. There is little evidence on crop yields. Georges Duby has argued that seed/yield ratios rose but his data have been trenchantly criticized by Slicher van Bath.[9] The adoption of the horse and the improved plough could have improved yields and labour productivity, but not until the later Middle Ages. Indeed the claim that there was an agricultural revolution between the sixth and ninth centuries, or between the ninth and twelfth centuries, seems a clear case of taking the introduction of an innovation rather than its majority adoption as a measure of progress.

The agricultural revolution in England

The agricultural revolution in England has been discussed by historians for over a hundred years. The earliest writers, such as Karl Marx and Arnold Toynbee, wrote however of an *agrarian* revolution; they emphasized the importance of enclosure and the concentration of landownership which allowed the adoption of new farming methods. But the first widely accepted model of the English agricultural revolution was that put forward by Lord Ernle.[10] Like Marx and Toynbee he believed that the adoption of new farming methods was impossible as long as the open fields and common land survived. Like Toynbee he believed that 1760 was the critical date. After then parliamentary enclosure rapidly extinguished the restrictions and inefficiencies of the common field system. Farmers, largely at the prompting of progressive landlords, particularly those in Norfolk, adopted new methods: these included the growth of turnips on the fallow, the use of the drill, more frequent hoeing to eliminate weeds, and the growth of clover and other legumes, which gave extra grazing and increased the nitrogen content of the soil. These crops provided extra livestock fodder, which increased the number of livestock which could be kept upon a farm, and in turn the supply of manure; manure supplies were increased by the stall-feeding of cattle, which were fed not only turnips and hay, but after 1800 imported oil-seeds. This extra manure increased the yields of cereal crops grown in rotation with turnips and clover in the celebrated Norfolk four course. The breeding of sheep and cattle was greatly advanced by following the methods of Robert Bakewell, and the quality and yield of milk, meat and wool improved. The plough – and other implements – were improved and so the creation of a good tilth in the seedbed, and the elimination of weeds, became more efficient. Between 1760 and 1815 English agriculture was transformed. It was this period that Ernle's interpreters believed was the period of the English agricultural revolution, although Ernle himself hardly used that term, and indeed wrote: 'The gigantic advances of agriculture in the nineteenth century dwarf into insignificance any previous rate of progress'.[11]

Ernle's views have been subject to much criticism by later writers. Some have doubted the importance of parliamentary enclosure; it has been shown that turnips and clover could be grown upon the open fields; that enclosure did not necessarily lead to improvement; and that much of the country had been long enclosed in 1760.[12] Nor were the new methods suitable to all areas. Turnips were difficult to grow upon heavy clay soils, which did not have their agricultural revolutions until efficient underdrainage was adopted after the 1840s.[13] Others have noted that even in Norfolk, the area on which Lord Ernle relied for much of his evidence, the adoption of new techniques was very slow; the drill was hardly used for turnips as late as 1800, and the Norfolk four course, the centre of Ernle's system, was not in general use upon the Holkham estate of the Cokes, its alleged classical *locus*, until the 1830s.[14]

But the most trenchant revision of the Ernle model came from Eric Kerridge.[15] His major criticisms were two. First Ernle neglected some important means of increasing productivity, notably the use of 'floating' meadows to provide fodder and the spread of convertible husbandry. In the open field system land in arable was confined to the growth of crops with alternate fallows; land in grass always remained in grass. Under convertible husbandry, land was sown to crops for a number of years; it was then sown to grass, later a mixture of grass and legumes such as clover, rye-grass, lucerne and sainfoin. This provided grazing and hay, rested the land and encouraged nitrogen fixation in the soil. The land was then again sown to crops. Convertible husbandry, argued Kerridge, led to pronounced increases in crop yields, particularly in midland England. Second, Kerridge argued, Ernle's chronology was quite erroneous. Far from the agricultural revolution beginning in 1760, improvements were adopted from the 1580s, spread rapidly in the seventeenth century, and were completed by 1767. After then there was little progress.

Kerridge's views have been much debated. Few agree that there was no progress after 1767, but a majority now agree that there was already much progress before 1760, Ernle's starting point. It would be generally accepted that the adoption of clover and turnips began in the mid seventeenth century, and was already widespread by the mid eighteenth century.[16]

Thus the general tenor of recent writing is to push the beginnings of the English agricultural revolution back into the seventeenth century, and perhaps even to the late sixteenth century. Some have also noted that nineteenth-century advances differed in kind from those of the eighteenth century. This has prompted F. M. L. Thompson to distinguish three separate agricultural revolutions. The first, or traditional agricultural revolution, was characterized by the extinction of the common fields, improved rotations and selective livestock breeding; it was completed by 1815. In the 1820s the second agricultural revolution began; it was characterized by the large scale purchase of inputs from off the farm, notably feeding stuffs and fertilizers, by field drainage and the construction of new farm buildings. The most rapid of these changes took place between 1820 and 1880. A third agricultural revolution, the adoption of labour-saving machinery, he dates from 1914.[17]

Although these writers have come to very different conclusions about the nature of the English agricultural revolution, they have all been concerned primarily with the adoption of new methods, less with their effect upon output, yields or productivity. To these we now turn.

The growth of output

Although there have been several attempts to estimate the growth of English agricultural output, they are all limited by the absence of any reliable figures before 1866.[18] In that year the Board of Agriculture began the annual publication of information on labour force, land use and livestock but not, until the 1880s, on crop yields or farm size. At no time has output been recorded in these returns, although two separate censuses of output have been held, and the Ministry of Agriculture has calculated trends in output since the 1950s. Thus any measures of output rely upon contemporary estimates of land use and livestock numbers. Some accuracy can be allowed to those estimates made in mid Victorian times and during the Napoleonic Wars; before then there is nothing but the estimates of land use made by Gregory King – and later modified by Charles Davenant – for the late seventeenth century.

Most historians have emphasized the importance of increased crop yields in the eighteenth and nineteenth centuries. They have thus neglected the very considerable increase in the area under cultivation which took place, particularly after the 1770s. Between 1700 and 1800 the arable area increased by 1.42 million ha, and by the 1850s by a further 0.7 million ha, an overall increase of some 62 per cent (Table 27). During this period the area in fallow declined, so that the *sown area* increased more rapidly – it nearly doubled – than the arable area; some of this increase was due to the expansion of cereal growing, which continued down to the 1840s; but there was also of course a great increase in the cultivation of fodder crops; by the 1850s one third of English arable was in roots and rotational grasses.[19]

The Board of Agriculture did not collect estimates of crop yields until 1886. Before then the evidence on trends in crop yields is sporadic and comes from different parts of the country; reliance has to be put upon contemporary estimates, which were often only for wheat; estimates of trends in livestock productivity are few before the twentieth century. Nor is it possible to estimate increases in crop yields from the rate of adoption of new techniques. Crop yields were a result of changes not only in inputs but in the weather and variations in plant and animal disease; it is impossible to estimate the precise effect of farm improvements upon yield, because of the complex interlocking of methods, inputs, soil type, weather and disease. There is no doubt that as early as the 1620s

Table 27 *Land use in England and Wales, 1696–1866 (million ha)*

	Arable	Fallow	Sown area	Pasture and meadow	Total cultivated area
1696*	3.64	1.09	2.55	4.86	8.5
1696†	3.64	0.53	3.11	4.86	8.5
1801	5.06	0.81	4.25	–	–
1854	5.91	0.36	5.55	5.26	11.17
1866	5.79	0.32	5.47	4.17	9.96

*Assuming one third of the 3.24 million ha in cereals, peas and beans was in fallow.

†Assuming one third or half the 3.24 million ha in cereals, peas and beans was in fallow.

Sources: C. Davenant, *The Political and Commercial Works* (London, 1771), vol. 2, p. 216; H. Prince, 'England *circa* 1800', in H. C. Darby (ed.), *A New Historical Geography of England* (Cambridge, 1973), pp. 403, 417; L. Drescher, 'The development of agricultural production in Great Britain and Ireland from the early nineteenth century', *Manchester School of Economics and Social Studies*, vol. 23 (1955), p. 167; Ministry of Agriculture, *A Century of Agricultural Statistics: Great Britain 1866–1966* (London, 1968) p. 94.

some farmers were obtaining high yields by the use of the best methods and heavy labour inputs; what is at issue is the trend of national yield.[20] A critical review of the evidence suggests that the average yield of wheat in England at the end of the seventeenth century was about 1009 kg/ha; no figure exists before this date, or for the mid eighteenth century, but by the early nineteenth century it was 1412 kg/ha, and in 1851 1691 kg/ha; it had reached 2018 kg/ha by the 1870s and was still at that level in the first decade of the twentieth century (Table 28).

These national estimates are confirmed by data for Norfolk and Suffolk; estimates of wheat yields from a sample of probate inventories suggest that the yield was about 538–673 kg/ha in the late sixteenth century, and 940–1076 kg/ha by 1700, most of this increase being achieved between 1600 and 1660; by 1801 the yield had risen to 1345 kg/ha and by 1850 to 2132 kg/ha (Figure 29). The best estimate then available is that English yields rose by about 75 per cent between 1700 and 1850. If the figures for Norfolk and Suffolk are correct, then there was an increase of about 66 per cent in the first two thirds of the seventeenth century, before the beginnings of clover and turnip adoption. This would bear out Eric

Kerridge's contention of an agricultural revolution beginning in the 1580s.

It is clearly not possible to estimate the growth of English agricultural output merely from the growth of the arable and sown areas and the rise in wheat yields. It gives us no guide to trends in livestock output; much of the increased sown area was devoted to fodder and not food crops. But assuming wheat yields to be representative of trends in all arable output then the total output by volume of all crops would have risen 264 per cent between 1700 and 1850. The increase in cereal output was probably less than this at about 170%.[21]

Changes in productivity

There is no doubt that there was a considerable increase in English arable output between 1700 and 1850, and no doubt that wheat

Table 28   *Estimates of crop yields, 1590–1905*

England and Wales		Norfolk and Suffolk	
Year	Yield (kg/ha)	Year	Yield (kg/ha)
		1590	538
1665	1009–1076	1665	941–1076
1700	807–1009	1700	941–1076
		1735	941–1076
1801	1412	1801	1345
		1836–8	1801
1850	1691	1850	2132
1861	1909	1861	2020
1870	1915	1870	2020
1885–90	1914		
1900–05	2051		

Sources: R. Lennard, 'English agriculture under Charles II: the evidence of the Royal Society's "Enquiries"', *Economic History Review*, vol. 4 (1951–2), p. 41; C. Davenant, *The Political and Commercial Works*, vol. 2 (London, 1771), pp. 216–18; P. G. Craigie, 'Statistics of agricultural production', *Statistical Journal*, vol. 46 (1883), p. 21; M. Overton, 'Estimating crop yields from probate inventories: an example from East Anglia, 1585–1735', *Journal of Economic History*, vol. 39 (1979), pp. 363–78; J. Fletcher, 'Contributions to the agricultural statistics of the Eastern counties', *Statistical Journals*, vol. 6 (1843), pp. 130–2; Ministry of Agriculture, *A Century of Agricultural Statistics: Great Britain 1866–1966* (London, 1968), pp. 108–9.

Table 29   *The labour input in English agriculture, 1700–1851 (millions)*

	Total population	Rural* population	Agricultural labour force
1700	5.8	4.8	1.2 to 1.4
1750	5.8 to 6.5	4.4 to 5.0	1.35
1801	9.2	5.8	1.7
1851	17.9	8.2	2.1

*In places of less than 2500.

Sources: D. Grigg, *Population Growth and Agrarian Change: an Historical Perspective* (Cambridge, 1980), pp. 95, 169; S. Pollard, 'Labour in Britain', in P. Mathias and M. M. Postan (eds.), *The Cambridge Economic History of Europe*; vol. 7, *The Industrial Economies: Capital, Labour and Enterprise*; part 1, *Britain, France, Germany and Scandinavia* (Cambridge, 1978), p. 141.

yields increased significantly. Nonetheless 62 per cent of the estimated increase in output can be attributed to the expansion of the arable area and the reduction of the fallow; only a third can be attributed to increases in crop yields. The growth in output also needs to be related to increases in the total population and to increases in the labour and capital inputs of the period. The total population of England and Wales rose by 200 per cent between 1700 and 1850, so the volume of arable output, which rose by 264

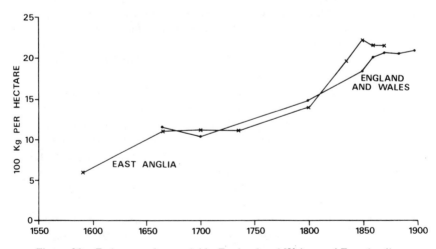

Figure 29   *Estimates of crop yields, England and Wales, and East Anglia*
Source: see Table 28.

per cent, exceeded population increase. It is less clear that food output as represented by the cereal output increase of 170 per cent did exceed population growth.

Changes in the labour input in English agriculture over this period remain a matter of controversy. Nineteenth-century writers believed that the agricultural population declined in the eighteenth century as parliamentary enclosure drove the small landowner to the towns, and improved technology replaced the farm labourer. This view is no longer held. There is no doubt that the rural population of England increased between 1700 and 1850; there was no rural depopulation until after the 1850s. Further, there was little if any use of labour-saving machinery before the 1820s and 1830s, while the reclamation of new land and the adoption of labour-intensive practices in arable farming probably increased the demand for labour.[22] Nonetheless some authorities still believe that the agricultural labour force was constant between 1700 and 1851.[23] Between 1700 and 1750 there probably was little increase in the labour input, but after 1750 the greater natural increase of the population suggests that at least the number seeking work in agriculture greatly increased, a view borne out by the unemployment and underemployment apparent in the 1820s and 1830s. Unfortunately there is little reliable information upon the number in the work force from 1700. The first reliable account is in the census of 1851; in earlier censuses there is some doubt as to the allocation of 'labourers' to agriculture and to non-agricultural activities and Gregory King's figures suffer from the same defect. Other sources suggest that there was an increase in the labour input; this increase was slight before 1750, not very great from 1750 to 1801, but more marked from 1801 to 1851 (Table 29). Over the whole period 1700–1850 the rural population increased by 70 per cent, and the agricultural labour force by between 50 and 75 per cent (Table 29). It is likely that these figures *understate* the increase; nonetheless it is clear that labour productivity rose very considerably between 1700 and 1850. The volume of output in the mid nineteenth century was over three times what it was in 1700, but the labour input increased only between 50 and 75 per cent. Arable output per head approximately doubled in this period.

These crude calculations confirm that there was a considerable increase in the output of English agriculture between 1700 and 1850; that there was an increase in output per hectare, and that labour productivity also rose. What they do not tell us is whether this productivity increase had been going on before 1700. The trend in yields before 1700, particularly between 1600 and 1665, suggests that Kerridge is right; if this is so then the English agricultural revolution will have to be revisited and rewritten. The implications for both the history of population and of industrialization are considerable. However it is worth noting how slow these changes were compared with the modern period. The volume of arable output grew at 0.8 per cent per annum in 1700–1850, compared with 2–3 per cent in Western Europe in the 1950s and 1960s. Wheat yields rose at 0.4 per cent in 1700–1850, and probably doubled between 1650 and 1850. Wheat yields in England doubled between in 1930s and 1960s. Labour productivity rose at approximately 0.5 per cent per annum in 1700–1850, again well below modern rates.

Capital inputs

The measurement of capital formation in agriculture presents severe methodological problems in modern times, when some data are available. It is a daunting prospect for the historian.[24] An attempt has been made to estimate fixed capital formation in Britain from the 1760s to the 1850s; fixed capital is defined as outlays on farm buildings, enclosures, reclamation, drainage and other improvements to the land, farm roads, equipment and machinery. The figure has been derived first by estimating national rent by decades, and second by assuming *expenditure* for these figures to be the same proportion as that found in a sample of estate records covering the period. At constant prices investment nearly doubled from 1761–70 to 1801–10, rose by 70 per cent in the next half century, and more than tripled over the whole century (Table 30). Assuming that there was some investment before the 1750s then capital investment rose more than arable output in 1700–1850. This is not surprising; the cost of drainage, reclamation and enclosure was formidable, and its benefits carried on to the present day.

The sources of productivity growth

Historians have been apt to attribute most of the growth of output and productivity in the eighteenth and nineteenth centuries to technological advance, and this can hardly be denied. Increases in

Table 30   *Fixed capital formation in agriculture in Great Britain, 1761–1860*

	Gross rent (£ million per annum)	Capital expenditure as a percentage of gross rent	Capital expenditure at current prices (£ million per annum)	Capital expenditure at 1851–60 prices (£ million per annum)
1761–70	20	6	1.20	2.18
1771–80	21	7	1.47	2.62
1781–90	24	8	1.92	3.31
1791–1800	31	11	3.41	4.26
1801–10	32	16	5.12	4.06
1811–20	41	14	5.74	4.45
1821–30	40	10	4.00	4.08
1831–40	42	11	4.62	4.71
1841–50	46	13	5.98	6.16
1851–60	46	15	6.90	6.90

Source: C. H. Feinstein, 'Capital formation in Britain', in P. Mathias and M. M. Postan (eds.), *The Cambridge Economic History of Europe; vol. 7, The Industrial Economies: Capital, Labour and Enterprise*, part 1, *Britain, France, Germany and Scandinavia* (Cambridge, 1978), p. 49.

the use of manures and fertilizers increased the plant nutrients in the soil as did the growth of pulses and fodder legumes; improvements in underdrainage and the use of marl and lime made these nutrients more available to plants. The use of better and new implements – such as the drill – allowed more frequent cultivations, and reduced competition from weeds. The steady replacement of the ox by the horse and the use of better implements increased labour productivity, although it was not until after the 1820s that there was much use of explicitly labour-saving devices, such as the thresher for the flail, and the scythe for the sickle, while the use of the reaper was delayed until after 1850.

However there are other explanations of increasing productivity. First the quality of physical inputs was continuously being improved; this was most noticeable in farm implements, whose design and construction improved slowly in the eighteenth century, and more dramatically in the nineteenth century, when factory-made implements became widely available. Other inputs improved in quality. The turnip was susceptible to disease, and was supplemented by swedes and mangold-wurzels in the late eighteenth century. Rotations were also improved; land cultivated with the Norfolk four course suffered from 'clover-sickness', and rotations were modified to prevent this. There probably were improvements in the quality of the seed sown, and in the quality of livestock, due not only to better breeding but better feeding. It may be that the quality of the labour force also improved. The supply of information to farmers increased (see pp. 159–60), particularly in the nineteenth century. It is possible that the reduction of infectious disease – although this has been much debated – produced a healthier labour force. One is hard pressed to argue that the labour force became better fed and thus stronger. Labourers' wages were buoyant in the early eighteenth century, but real wages declined after 1750, and did not rise again until the 1820s, possibly until the 1840s. It has been argued that the differences in agricultural productivity between the south and north in the nineteenth century were due to the low wages, poor diet and hence physical debility to be found in the south; in the north industrial competition gave higher agricultural wages and a healthier, fitter population.[25]

The eighteenth and nineteenth centuries may have seen some productivity increases due to regional specialization, as transport conditions improved, and allowed some areas to specialize in those goods in which they had a comparative advantage. The slow concentration of cereal growing upon light soils, and the tendency for lowland clay vales to turn to livestock production might illustrate this. Certainly there must have been some gains from the slow movement towards larger farms. From the earliest eighteenth century there was an undoubted tendency to amalgamate, so that an increasing proportion of England was farmed in large units. This allowed economies of scale to be exploited, although it may be that this was delayed until the advent of labour-saving machinery.[26] Certainly the consolidation of the fragmented holdings of the open fields into compact farms must have had consequences for productivity for it reduced the time spent on movement, and made the control of animal and plant disease easier.

Conclusions

It is possible to show that the eighteenth and early nineteenth centuries were a period of growing output and rising labour and land productivity. Unfortunately it is less easy to be precise about when this began. It is not possible to measure any trends before

1700 except for yields in East Anglia; but it is perfectly possible that rates of increase comparable to the eighteenth and nineteenth centuries were being obtained then. Nor can much be said about changes in the rate of increase *within* the period 1700–1850, although there is a suspicion that the first half of the nineteenth century saw a greater rate of change than the eighteenth century. The absence of any reliable data makes it impossible to confirm the traditional view of a spurt from the 1750s, although the Norfolk yield data might suggest this. It is perhaps however worth pursuing approaches such as those demonstrated in this and the preceding two chapters, as well as continuing in traditional ways.

Part Five
Institutions and agricultural change

14 Class, region and revolution

So far a series of models have been dealt with that explore the impact of population growth, environment, industrialization, transport and diffusion upon agrarian changes, while some time has been spent considering what is meant by the concept of agricultural revolution. We turn finally to consider two models of change in society that touch upon agriculture but are not concerned only with agriculture. Both deal with what can be conveniently described as institutions and their role in economic and social change. In Marx's work the engine of change is the conflict between classes; in the specific case that is considered, his view of English economic change between the fifteenth and nineteenth centuries, the classes are defined in terms of the control of the means of production, and particularly the ownership of land. The institutional process that separated men from the land and created the characteristics of nineteenth-century agrarian capitalism was enclosure.

F. J. Turner's views on the frontier have been applied to many countries, but he was concerned with the way in which the frontier – the advancing edge of agricultural settlement – gave rise to particular economic, social and political forces that gave American life distinctive institutions. Thus he was in a sense concerned not with how institutions influence agrarian change, but how agrarian change influences institutions. Nor does the work of Marx and Turner exhaust the possible interrelationships between institutions and agrarian change. The concept of property rights has provided the basis for a recent theory of economic change, while other examples would include the numerous studies of plantations, slavery and agrarian change.[1]

A review of the very large literature on the problems of the underdeveloped world will reveal two contrasting approaches. To some the problems are to be solved by technological advances, and they discuss the type of new farming methods necessary and the ways in which farmers may be persuaded to adopt them. Other writers believe that land tenure is the heart of the problem, and that it is only land reform that can lead not only to greater productivity but to greater social justice.

The latter approach can be traced back to, among others, the writings of Karl Marx. Marx's work was primarily concerned with criticizing the capitalist economy of his time, and attacking the bourgeois economists who supported the system. But he also wished to show how the capitalist economy had come into being, and how it would collapse, due to its own internal contradictions, being replaced by socialism. Scattered through his works are several sketches of a theory of historical change, and in volume one of *Capital* there is an account of changes in the English economy from the fifteenth century. Although this is not specifically a model of agrarian change, it shows how a country changes from an agrarian, feudal society to an industrial, capitalist society.

In his early work Marx argued that all societies went through the same sequence of economic stages from primitive communalism via the ancient city state based upon slavery, feudalism based upon serfdom, capitalism based on free wage labour, followed by communism. He wrote little upon any of these stages save capitalism, and the transition from feudalism to capitalism.

His notebooks for *A Critique of Political Economy*, not published in English until 1972, under the title *Grundrisse*, contain a fuller account of the earlier 'modes of production'. In this the idea of unilinear evolution is less emphasized, and indeed some modern Marxists put less and less emphasis upon this sequence.[2]

Marx, like other classical economists, thought labour was the only source of wealth. Every labourer needed to work in order to maintain and reproduce himself. All value produced above this maintenance level was surplus value. The distinguishing feature of any 'mode of production', his term for a stage of development, was the way in which this surplus value was controlled; in each 'mode of production' there were particular social relations between the labourer, the ruling class that appropriates the surplus value, and the means of production. Each 'mode of production' contains inherent contradictions; conflict between the two prominent classes leads to crisis in the system, which changes inevitably into the next system. Thus in feudalism lord and peasant clash, and from this emerges the bourgeoisie and the industrial proletariat, who are in conflict within the capitalist mode of production, which in turn collapses with the triumph of the proletariat and the emergence of communism.

Historians and others have attempted to apply Marx's analysis to a variety of places and periods; but his fullest account of the rise of capitalism and the changes in agrarian society which are necessary for this to come about refer only to England; it is this description that is outlined here. Since he wrote, his ideas have been amplified and criticized by many others, and so this account incorporates later work by Marxists, rather than relying solely upon Marx's work.

Marx had little to say on the feudal economy, and the descriptions of the feudal mode of production and its decay are mainly the work of modern Marxist writers.[3] This chapter deals mainly with Marx's view of the evolution of English agriculture from the fifteenth to the nineteenth century.

The emergence of an agrarian capitalist class and a landless proletariat

Marx deals with the period in England between 1450 and the mid nineteenth century in his chapters on the So-called Primitive Accumulation in volume one of *Capital.* He attempted to show how the capitalist mode of production grew from the decay of feudal society. The central theme is how the peasants of the fifteenth century, who controlled the means of production, land, tools and livestock were separated from the land and became labourers, without control of the means of production, and therefore having to work for wages. He also tries to account for the rise of a landlord class, who owned progressively more of the land, at the expense of the peasantry, who had disappeared by 1750, and the emergence of a class of capitalist tenant farmers, who rented land from the landlord class and hired landless labourers to work it. Further, the farms that these tenants rented grew progressively larger, so that the small farmer sank to the level of a landless labourer. In the eighteenth and nineteenth centuries the rural landless provided the labour for the new factory industries of the towns.[4]

The agency, or institution, that separated the peasant from the land was enclosure, which was imposed by force upon the peasant in the sixteenth and seventeenth centuries, but backed by the law thereafter. This process began between 1465 and 1525 when the end of feudalism and the disbanding of the feudal retinues created a landless class. A new aristocracy, more concerned with cash than war or display, enclosed villages for sheep grazing, further increasing the landless. Attempts by the state to prevent depopulation proved fruitless. After the reformation the sale of ecclesiastical land led to further evictions as new landlords drove out subtenants and created large farms. The Civil War was a further turning point, for Crown lands were sequestered and sold and the new bourgeois Parliament from then on supported large landed property owners, and expropriation of the land was backed by the law. By 1750 the yeomanry had disappeared, and by the 1790s the common lands were extinguished, depriving the cottager of his last support, and driving him to the industrial towns. In the late eighteenth and nineteenth centuries capitalistic agriculture emerged, based upon the landless labourer, the capitalist tenant farmers, and the rentier landlord. Farms became progressively larger, and scientific agricultural methods were adopted. Enclosure and the adoption of scientific farming methods led to a reduction in the demand for labour, a fall in the labourer's wages, and a decline in the agricultural population. There was a continued emigration from the land; by the 1850s there was a stark contrast between the efficiency of capitalist agriculture and the deplorable state of the labourer, exemplified by conditions in England's leading agricultural county: 'The cleanly weeded land, and the uncleanly human weeds of Lincolnshire are pole and counterpole of capitalist production'.[5]

Revisions of Marx's ideas

Since Marx there has of course been much written upon the topics he raised in his outline of change from 1450 to 1850. But whatever the views held by historians upon these issues, it has proved remarkably difficult to confirm or refute his arguments for a number of reasons. First, and perhaps most important, is the absence of reliable national statistics upon landownership, farm size or enclosure before the nineteenth century. The only national returns of landownership made in England were those of 1873–4, which record the areas owned by landlords in every county in Britain. Prior to that changes in landownership can only be traced in archival material representative only of a very small sample of landowners, often in limited areas. Thus, for example, H. J.

Habakkuk's important paper on landownership in England in the late seventeenth and early eighteenth centuries is based upon only two counties.[6] It is even more difficult to trace changes in farm size. The censuses of 1831 and 1851 collected some data on farm size, but the Ministry of Agriculture has only collected and published such data since 1885. Before then information on changes in farm size has to be drawn from occasional surveys of parishes or estates; few of these have been analysed and published. Indeed before the eighteenth century there is very little information on trends in farm size other than the polemical remarks of pamphleteers. Further complications arose in a failure to distinguish between consolidation, where the scattered strips of one farm are grouped together to form a compact holding, and amalgamation, where two or more farms are united as one farm. The term engrossing has also been used and interpreted in different ways.

Enclosure is also poorly recorded, for the only national record of enclosure is in the parliamentary Acts and Awards. It is not always clear whether the areas referred to in the Acts are waste or common field, and in some cases the Acts merely confirm enclosures made in the past. Enclosure awards are a more authoritative record of areas of waste and common field enclosed by Act – that is to say mainly after 1740. Before then there is no comprehensive account of how much land was enclosed or where, save in records which clearly miss much enclosure activity. Thus fifteenth- and sixteenth-century Acts which made it illegal to cause the decay of arable husbandry led to royal commissioners taking evidence with a view to prosecution; but the statistics gathered for this evidence have been questioned, and are far from comprehensive. Some of the many enclosures by agreement were ratified by Chancery decree, and these records provided evidence, again incomplete, on seventeenth-century enclosure.[7]

A second difficulty in assessing trends in enclosure, landownership and farm size is the failure to specify exactly what was changing. Thus Marx and subsequent writers wrote of the decline of the English peasantry without defining peasantry. However the essential features would seem to be that the peasant controls, although not necessarily owns, the means of production, that is land, implements and livestock; that his land is worked mainly with family labour and without the use of permanent hired labour; and that he aims to provide for much of his family's needs from his own land, therefore selling a small proportion of total output. Changes in the pattern of

English landownership have also been confused by the use of the terms the gentry, and the yeomanry, neither very clearly specified. A further complication arises from the failure to distinguish between units of production or farms, and units of ownership, or estates; in many descriptions of the 'engrossing' that was thought to accompany enclosure it is not clear whether this refers to the creation of bigger farms or bigger estates. Finally, while Marx and many subsequent writers have referred to the growth of large farms at the expense of small farms, small and large are rarely defined.

The chronology and chorology of enclosure

When Marx wrote – and indeed until the late nineteenth century – it was thought that open fields had once existed throughout England and Wales, and that they had been extinguished in two periods of enclosure: in the sixteenth century and in the century between 1750 and 1850. It is now believed that as early as 1500 open fields were uncommon in many parts of the country, either because they had never existed, or because they had already been enclosed. Thus in 1500 little open field arable remained to be enclosed in Wales, in Cornwall, in east and south Devon or western Somerset; most of the open fields had gone in the Welsh borders, from Cheshire south to Monmouth, and they were unusual in Lancashire. In the south east there was already a zone with few surviving open fields, from eastern Norfolk and Suffolk into Essex and Kent. There are few records of how and when this enclosure took place. As long as waste land was abundant, as it was in the fourteenth and fifteenth centuries, individual assarts could be hedged without much protest from others. Probably while population densities were low, as they were after the Black Death, there was little objection to individuals exchanging strips to form compact farms, for there was little shortage of grazing. Where landlords found themselves in sole possession of a village after the Black Death, enclosure could have taken place without opposition.

But as soon as population began to increase again, there was opposition to enclosure. From the 1480s and throughout the sixteenth century pamphleteers wrote on the injustices of enclosure. It was argued that the rise of wool prices led landlords to consolidate strips, to evict tenants and to amalgamate their holdings into larger farms. In addition the conversion of arable land into pasture reduced the employment available. This, it was said, led to the depopulation of whole villages. The arguments of the pamphleteers were accepted by R. H. Tawney in his book *The Agrarian Problem of the Sixteenth Century*.[8] He agreed that there was little enclosure in the west, north west and south east, and that most of the enclosure took place in a number of midland counties. He thought however that E. F. Gay's estimate that only 202,000 ha of cultivated land were enclosed between 1450 and 1607, causing the eviction or unemployment of from 30,000 to 50,000, was an underestimate. Gay's figures were based upon government inquiries into the decline of tillage made in 1517–19, 1548, and 1607. The figures for 1517–19 cover only twenty-three counties, those for 1607 only six. Gay assumed that the areas recorded in 1517–19 covered the areas enclosed from 1485, and then assumed that a similar rate of enclosure had held in 1450–85, 1518–77 and 1578–1607. Clearly his estimate is dubious, and he concluded that enclosure affected only 2.76 per cent of the total land area of twenty-four counties in a century and a half.[9] Even in the midland counties most affected, no more than a tenth of the total area was enclosed (Figure 30). Furthermore, wholesale enclosure of a parish was rare; in a majority of cases comparatively small areas – less than 40 ha – were enclosed. Subsequent work on enclosure in the fifteenth century suggests that the rate of enclosure before 1485 was perhaps greater than Gay believed and that there was comparatively little enclosure for conversion to grass after the 1520s.[10] There was certainly however a continuation of piecemeal enclosure, which aroused little outcry. By 1600 the pattern of enclosure reinforced the distinction between the south east and the west, where the open fields were uncommon, and the wedge of land running from the East Riding south westwards to Wiltshire, where much of the arable was still in open fields. The source for E. C. K. Gonner's map of proportion enclosed in 1600 is unclear; while the proportions cannot be accepted in absolute terms, the relative distribution accords with most accounts by modern writers.[11]

Early historians believed there was little enclosure in the seventeenth century, if only because there is little record of it. Enclosure of commons in the Fens aroused a further outburst of pamphleteering, while it doubtless proceeded quietly in upland areas where there was abundant grazing. There was also general enclosure of open fields by private agreement, evidence of which survives in Chancery decrees, as well as much piecemeal enclosure

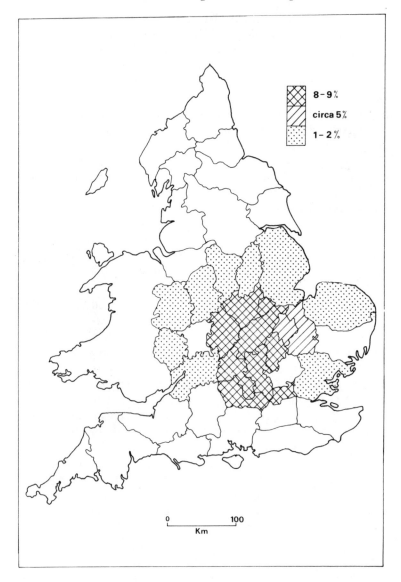

Figure 30 *Enclosure in England: percentage of total area enclosed 1450–1607*
Source: E. F. Gay, 'Enclosures in England in the sixteenth century', *Quarterly Journal of Economics*, vol. 17 (1903), pp. 576–7.

by agreement which has gone unrecorded. It is not until the eighteenth century that documentary evidence of areas enclosed is available, for parliamentary Acts and enclosure Awards contain figures for the area of waste and common land enclosed by Act, from 1700 to 1914. E. C. K. Gonner and G. Slater both produced maps showing the distribution of enclosure, both showing that most of the enclosure of common field took place mainly in the midland zone.[12] However their estimates of area are considered defective. The most authoritative figures are those collected by W. E. Tate, who distinguished the area enclosed from common grazing and waste alone, and the Acts or Awards which enclosed land including some common field arable. M. E. Turner has collated these figures and concludes that 2.7 million ha of land were enclosed by Act, 20.9 per cent of the total land area of England and Wales.[13] Of this, Acts that enclosed land including *some* common arable land accounted for 1.78 million ha, while Acts referring only to waste accounted for 0.93 million ha. Unfortunately it has been shown that Tate's figures for Acts enclosing land including some arable greatly exaggerate the area of arable enclosed by Act; that is to say we cannot assume that the figure of 1.78 million ha is the area of arable remaining to be enclosed in 1700.[14] Furthermore there is no doubt that some Acts merely confirmed enclosures that had taken place in the past, while some authorities such as G. Slater believe that there was considerable enclosure in the eighteenth century by private agreement. Nonetheless Tate's figures are the most reliable available (Figure 31). By 1700 the open fields had disappeared from the south west; from the counties south of London from Dorset to Kent; from east Norfolk, Suffolk and Essex; from Wales and the Welsh border, from Cheshire, Stafford and Lancashire, the Lake District, and from most of the West and North Ridings, Northumberland and Durham. The places where common field arable was enclosed by Act lay in a zone from the East Riding south through Nottinghamshire and Lincolnshire and south westwards into Wiltshire. Only in Oxfordshire was common field enclosed by Act more than half the total land area of the county. Waste enclosed by Act was a significant proportion of the total land area only in the northern upland counties, the fenlands of Lincolnshire, Cambridgeshire and Somerset, and the mosses of Lancashire.

If Tate's figure of 1.78 million ha could be assumed to apply to open field arable, then in 1700 approximately half of the 3.64–4.05

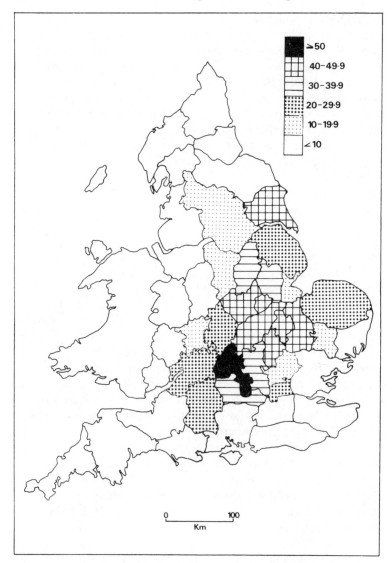

Figure 31   *Arable land enclosed by parliamentary Act in England: area*
including some commonfield arable that was enclosed by
parliamentary Act as a percentage of total area of each
county
Source: M. E. Turner, *English Parliamentary Enclosure: Its*
Historical Geography and Economic History (Folkestone,
1980), pp. 178–9.

million ha which were thought by King and Davenant to be in arable in *c.* 1690 were already enclosed by that year; but the remaining half was enclosed in a remarkably short time. 80 per cent of all parliamentary enclosure took place between 1760 and 1820. But there are good reasons for supposing that the figure should be lower. But whatever we conclude from this, there is no doubt that parliamentary enclosure affected a very limited region of England and only one fifth of its total area. It must be concluded that enclosure in England had taken place more slowly, more quietly, and more than is normally supposed by agreement.

The expropriation of the peasantry

Marx dated the real beginnings of capitalism in England from the sixteenth century, and thought 'the expropriation of the agricultural producer, of the peasant, from the soil, is the basis of the whole process'.[15] In the late fourteenth century, he thought, the bulk of the population were free peasant proprietors, with few labourers: the small group of independent labourers all had 1.6 ha of land and rights of common, while peasants who had smaller farms worked on large estates in their spare time. Between 1470 and the early nineteenth century the peasant lost his land, and became a free wage labourer; the independent labourers lost the rights of common; feudal landlords were replaced by a new and commercially minded class of landlords; and the capitalist farmer who rented land and hired wage labour emerged. Farms grew progressively larger.

Changes in the pattern of English landownership have puzzled many historians, and the lack of data has precluded any authoritative account. Equally perplexing have been the varying definitions and inconsistent application of the terms peasant, yeomen and gentry. Marx, for example, speaks of free peasant proprietors in the fifteenth century, but notes that by 1690 the 'yeomanry, the class of independent peasants', were more numerous than the class of farmers. By 1750, he wrote, the yeomanry had disappeared.[16]

If the distinctive feature of peasantry or yeomanry was that they owned the land that they cultivated then England was not a nation of peasant proprietors in the fifteenth century, for in 1436, J. P. Cooper has estimated, great landowners, the Crown and the Church owned some 40–45 per cent of the cultivated area. Small landowners with an income of £5–100 a year owned a quarter; not all of these latter would have farmed all the land they owned. F. M. L. Thompson has argued that in 1500, the upper limit owned by peasants – including copyholders – was half the cultivated area. On the basis of Gregory King's figures on freeholders and tenant farmers he estimated that the peasant owned 33 per cent of the cultivated land by *c.* 1690; the land tax returns suggest that the small landowner occupied 15–20 per cent of the cultivated area by 1790, and in 1870 owner-occupiers of less than 120 ha occupied 10 per cent of the cultivated area, if home farms are excluded.[17] However vague the bases of these estimates they support the belief that the area cultivated by farmers who owned their land declined from about a half in 1500 to a fifth or less in 1790. Recent historians believe that the critical period of decline was in the late seventeenth and early eighteenth century; there is little evidence that it was parliamentary enclosure that led to a radical decline in the numbers of small landowners.[18]

But owner-occupiership is not the only definition of a peasantry. The term implies also small scale production, largely for the family needs rather than for the market, and using primarily family labour. It is almost impossible to trace the decline of subsistence farming in England and no attempt will be made to do this. Figures on farm size are unreliable. However there is agreement among all historians that in these matters England was very different from the rest of Europe; at no time were farms subdivided and fragmented as they were in much of Western Europe, and the long term trend in England was towards an increase in the average size of farm.

Marx and his later followers, the latter apparently influenced by Lenin's work on the differentiation of the peasantry in Russia, believed that in the fifteenth century and earlier there was an approximate equality of farm size in England, with the bulk of all peasants having 6–8 ha, while there was another class of cottagers who had less than 2 ha but also rights of common grazing.[19] From the late fifteenth century this equality broke down, and English agriculture became increasingly characterized by the preponderance of large farms on the one hand, and landless cottagers who worked on the large farms for wages on the other hand. This perhaps neglects the existence of *demesne* farming, not only in the fifteenth century but earlier. In the thirteenth century much of the surplus grain and wool was produced on large manorial holdings worked by peasants. Nonetheless what little evidence there is on farm size

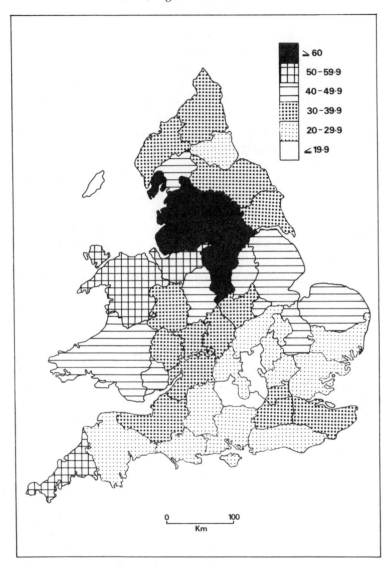

Figure 32   *Small farms in England and Wales, 1851: farms of 2 ha and over and under 20 ha as a percentage of all farms of 2 ha and over*
Source: *Census of Great Britain*; vol. 1, *Age, Civil Condition, Occupations and Birth Places: Accounts and Papers*, vol. 88, pt 1 (1952–3), pp. 30, 120–2, 236–340, 312–13, 406–8, 513–16; pt 2, pp. 596–8, 657–8, 728–9, 800–2, 884–5.

suggests the erosion of the small and middle sized farms in the sixteenth and seventeenth centuries when holdings of 6–12 ha became too small to provide an adequate livelihood, were vacated and amalgamated to become larger farms.[20]

This process became more marked in the eighteenth century when parliamentary enclosure, according to writers such as the Hammonds and H. Levy, gave landlords the opportunity to evict tenants of small holdings, which were thrown together in large farms, in much the same way as landlords evicted small farmers in the sixteenth century and created large sheep runs. However there is very little statistical evidence to support these arguments. Where estate surveys exist at different dates they invariably show a growth of the larger farm compared with others. But this was a slow process, and by no means confined to areas where parliamentary enclosure was taking place; nor were areas which were enclosed invariably the victims of amalgamation.[21]

There are no national statistics on farm size until the census of 1851. In that year small farms – farms of between 2 and 19.8 ha – were 41.1 per cent of all farms of 2 ha and more in England and Wales. Such farms were at least a fifth of all farms in all but two English counties (Figure 32). But they were most numerous in Cornwall, Wales, Lancashire and Cheshire, Westmorland, the West Riding, Staffordshire and Derbyshire all areas where most of the open fields had been enclosed by 1500. They were also more than 40 per cent of all farms in Nottinghamshire, Lincolnshire, Cambridgeshire and Norfolk. In the latter three counties later evidence on farm size suggests that the small farms were concentrated within the fenland parts of the counties. Thus by 1851 the small farm was far from being eliminated from the agriculture of England and Wales; and the large farm – over 120 ha – was important only in the east and the south.[22]

A third distinctive feature of a peasantry is the unimportance of a hired labour force; conversely the distinctive feature of capitalist agriculture is the existence of large farms worked by hired labourers. Recent work suggests that the landless labourer was more numerous in the thirteenth or fifteenth centuries than earlier historians believed. But there are no reliable statistics until 1851. In the census of 1851 farmers were asked to return the number of labourers they employed. On farms of 2 ha and over 40.6 per cent of all farmers employed no labour other than that of their wives and children. The regional pattern is, not surprisingly, like that of

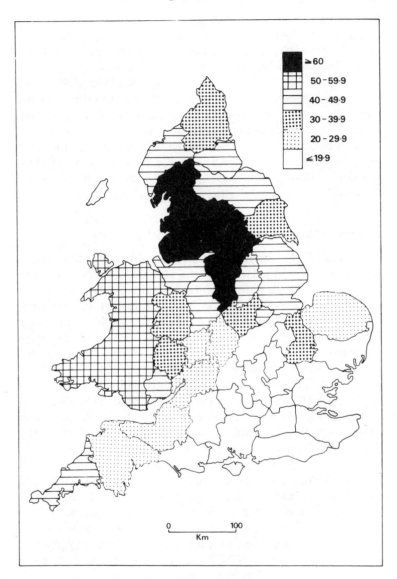

Figure 33   *Farms that employed no hired labour as a percentage of all*
farms of 2 ha and over, 1851
Source: *Census of Great Britain*; vol. 1, *Age, Civil Condition,*
Occupations and Birth Places: Accounts and Papers, vol. 88,
pt 1 (1952–3), pp. 596–8, 657–8, 728–9, 800–2, 884–5.

the distribution of small farms (Figure 33). In fact English farm labourers were very much concentrated on the large farms found on the uplands of the East Riding, Lincolnshire and in south east England south of a line from the Wash to Chesil Beach.

Thus while it is true that the period between 1450 and 1850 saw a decline in the freeholder or owner-occupier, the area they occupied falling from about one half to 15–20 per cent of the cultivated area, the processes of enclosure, amalgamation, engrossing and eviction did not end the small family farm in England; indeed in much of the south west, Wales and the north west this was still the dominant unit of production. In 1851 it was only in the south and the east that the capitalistic farm was dominant.

The rise of landlessness

Marx believed that in the fifteenth century there were few landless labourers in rural England, and the bulk of the population were peasant proprietors. Subsequently enclosure and eviction led to the creation of a landless class in the sixteenth century and then in the period of parliamentary enclosure after 1750. In this latter period the expropriated went to the towns to form the industrial proletariat.[23]

It has proved difficult to confirm or refute these views. It now seems likely that there were more landless in the fifteenth century than nineteenth-century writers believed; indeed the landless class – or at least a class with only a cottage and a hectare or so – were numerous in the thirteenth century. Most historians believe that the landless increased both absolutely and proportionally in the sixteenth century, and gave rise to the problem of vagrancy; but whereas Marxists would see this as largely due to enclosure, engrossing and the conversion of arable land to pasture, others believe it to be due to population growth. There are no reliable national figures, but village studies do show that landlessness was increasing. In Myddle, in Shropshire the landless were 7 per cent of the population in 1540–70, but by 1631–60 31.2 per cent, while in three villages in Arden, 40 per cent of the houses had no land in 1663–74. At the end of the seventeenth century Sir John Clapham estimated, modifying Gregory King's figures, that there were 330,000 farmers in England and Wales, and 573,000 labourers, some 63 per cent of the agricultural population. There are no comparable estimates for earlier periods, although it has been estimated by a recent historian that between 25 and 33 per cent of the farm population in the early Tudor periods were without land. This suggests that there was a substantial increase in the landless in the sixteenth and seventeenth centuries.[24]

But Marx's remarks on the creation of a landless population in the eighteenth and nineteenth centuries have attracted most attention. He and subsequent writers such as J. L. and B. Hammond argued that parliamentary enclosure was unfair to small landowners and to cottagers without land but with grazing rights. The small farmer could not afford the cost of enclosure and his land was bought by the greater landlords, while the cottagers lost grazing rights and thus their means of subsistence. Further, small farms were absorbed by large farms. The adoption of improved farming methods led to a reduced demand for labour. Thus the landless migrated to the towns where they provided the labour force for the new factories of the industrial revolution. In the nineteenth century the destruction of rural industries by factory products and the clearing of cottages in 'closed' parishes further contributed to the decline of the agricultural population.

This interpretation has been subject to much criticism. E. C. K. Gonner was unable to find any correspondence between enclosure and population growth, and J. D. Chambers showed that a number of agricultural villages in Nottinghamshire enclosed in the early nineteenth century continued to increase their populations.[25] Indeed few would now deny that the rural and agricultural populations of England increased substantially after 1750 (see p. 187).

It was not until after 1841 that any county in England showed an absolute decline, although after 1851 the agricultural labour force and the rural population declined until the end of the century. Thus it is not true that there was agricultural depopulation after 1750. Marx did not state why the adoption of new farming methods should have reduced the demand for labour; it may be that he believed that enclosure in the eighteenth century was undertaken in order to convert arable land to pasture, as it was thought to have been in the sixteenth century; one modern writer believes that this was so in the 1760s and 1770s,[26] but not afterwards when waste and common grazing were being ploughed up. Nor were the new farming methods specifically labour-saving. The growth of labour-intensive crops such as fodder roots and potatoes, and the increased frequency of weeding, the greater attention paid to livestock, the more frequent use of manure, marl and lime, the prodigal use of labour in stone-picking, hedging and ditching all suggest that there was an increase in the demand for labour. It was not until the agricultural labour force began to decline in the 1850s that the adoption of labour-saving machinery such as the threshing machine or the reaper made substantial progress.[27]

The consequences of parliamentary enclosure for the rural population have been much debated. It is agreed that the Hammonds overstated their case; those who suffered most were cottagers without established rights of commons; and probably some small landowners who could not pay the cost of enclosure. That these men suffered and provided a *potential* labour force for the towns is not in doubt. That there were major amalgamations of farms is more questionable. Recent work suggests that there was undoubtedly a trend towards large farms in the eighteenth and nineteenth centuries, but that it was not dependent upon enclosure, for it occurred in old enclosed areas as well as the areas of parliamentary enclosure.[28]

But was parliamentary enclosure the sole cause of the reserve army of labour for the new towns? There is no doubt that rural–urban migration was a major contributor to the urban growth of 1750–1850. The towns grew much more rapidly than the rural areas; indeed at a pace that could not be explained solely by natural increase. But recent opinion suggests that it was primarily rural natural increase rather than parliamentary enclosure that created the growth of landlessness. The rise of the rural population however was slow enough before 1790 to be absorbed in the new labour-intensive agriculture. During the Napoleonic Wars a substantial proportion of the population was engaged in the armed forces. But after the end of the wars 500,000 servicemen returned to the country. Rural population continued to grow by natural increase; but there was a marked difference between the experience of the agricultural populations in the north and the midlands on the one hand, and those in the south east on the other. In the south east there was underemployment and unemployment in the areas of large scale arable farming; in the north and the midlands the growing employment opportunities in the towns attracted the landless, and kept up wages among those who remained on the land. In the 1770s agricultural wages in the north and south had been much the same; but by the 1850s there was a marked difference. Thus rural–urban migration from the south east had been insufficient to reduce increase, and population grew beyond the opportunities for employment between 1820 and 1850.[29]

There are two reasons for supposing that parliamentary enclosure was not the main cause of rural–urban migration and not the provider of the bulk of the industrial labour force.

In the first place the majority of the migrants to British towns came from relatively short distances away. This was demonstrated by E. G. Ravenstein in 1885 and subsequent work has confirmed this. In 1851 for example, 70 per cent of the migrants into Preston had come less than 48 km. If then the industrial towns recruited most of their labour primarily from within 80 or 100 km, parliamentary enclosure cannot have provided labour for the industrial towns of Lancashire, for there was little or no parliamentary enclosure of open fields in that county[30] (see Figure 31). In the north east the enclosure of open field could have played no role in providing labour for the towns of Teeside and Tyneside, for the open fields had gone by 1750. Similarly only a tenth of the open fields of the West Riding were enclosed by parliamentary Act, while in South Wales enclosure had a limited role to play in causing migration in the later nineteenth century. The place where parliamentary enclosure of open fields could have been important in providing urban migrants was London, for a substantial proportion of common field was enclosed by parliamentary enclosure in the counties from Berkshire to Cambridgeshire; and the east midlands, where both Nottinghamshire and Leicestershire had much parliamentary enclosure.

Second, it is doubtful if the number of people made landless by parliamentary enclosure between 1750 and 1850 was large enough to account for all the rural–urban migration in that period, or indeed more than a small proportion of it. The population living in places of less than 2500 or more rose from 4,417,000 in 1751 to 5,820,000 in 1801, then to 8,239,000 in 1851, not quite doubling in a century, and growing more rapidly in the nineteenth century. The urban population however rose from 1,478,000 in 1751, and was 3,009,000 in 1801 and 9,687,000 in 1851.[31] A very crude estimate suggests that in 1751 no more than 600,000 people lived in areas which were subsequently to have their open fields enclosed. In 1831, when the enclosure of open fields was largely complete, some 1,300,000 lived in areas which had had their open fields extinguished in the preceding eighty years. As a very small fraction of those who lived in these areas were actually displaced by parliamentary enclosure, it seems hardly likely that they could have provided *all* the rural–urban migration that took place between 1750 and 1851. Thus the natural increase of the rural population, and its growth beyond the employment opportunities of the countryside, combined with the attractions of the new towns, seems a more plausible explanation of the movement than parliamentary enclosure.

Conclusions

When Marx wrote, comparatively little was known of the economic history of England. Subsequent work suggests that he greatly overstated the role of force in enclosure, misunderstood the trends in the rural and agricultural populations of the eighteenth and nineteenth centuries, and exaggerated the decline of the peasantry, not least because he was unaware of regional differences in farming structure. It is perhaps also unfortunate that his polemical style has carried on into modern debates. Nonetheless his work has stimulated a long debate on agrarian change; whereas he had little to say upon land use or technology, his emphasis on the role of landownership remains a valuable way of looking at agrarian change.

15 On the frontier

Much human history can be interpreted in terms of the migration of a people into a sparsely populated area. Thus the Bantu-speaking peoples moved slowly east and south from their homelands in West Africa in the first millennium AD, while the Han Chinese spread slowly but inevitably southwards at much the same time. In medieval Europe the Germans moved east of the Elbe in the twelfth and thirteenth centuries, while Muscovy expanded east and south until halted by the Mongols in the thirteenth century. But by far the greatest expansion has been that of the European peoples outwards into all the other continents from the sixteenth century. In Asia political dominance was not accompanied by settlement in any numbers. But in the seventeenth century the French and the English colonized the North American coastlands and the West Indies, while the Spanish and Portuguese settled in Central and South America, and the Dutch in South Africa. In the eighteenth and nineteenth centuries the British colonized Australia and New Zealand. In all these areas people of European origin slowly advanced into the interior of these continents subduing the indigenous peoples, exploiting mineral deposits and making agricultural settlements.

There has of course been much study of this expansion by historians. But in 1893 a young American historian, Frederick Jackson Turner, addressed an audience in Chicago on the significance of the frontier. In this lecture, subsequently published as an article, Turner argued that the characteristic features of American life were not derived from Europe, but from the experience of the frontier. This prompted a massive re-evaluation of American history, and much research on the spread of American settlement westwards; by the 1920s historians in other countries were comparing the European frontiers in other areas with that of the United States. Few essays have prompted such an enormous flood of literature.[1]

Turner's essay was not a model of agrarian change; its prime aim was to emphasize the influence of the frontier upon American institutions and attitudes, but his description of the westward move contained numerous provocative generalizations about the nature of agricultural colonization. It was not a rigorous model in the modern sense. Indeed it has been described as simply a series of propositions, largely unsubstantiated at the time. Nor did Turner make much effort to pursue his ideas in later life. Richard Hofstadter has remarked that Turner's reputation, like that of Lord Acton, depended upon the books he did not write.[2]

The significance of the frontier in history: a summary

Turner argued that 'until 1890 American history has been in large degree the history of the colonization of the Great West. The existence of free land, its continuous recession and the advance of American settlement, explain American development'. In 1890, according to the author of the census report, the frontier was closed. There was no longer a large area with a density of less than 5 people per km² into which settlement could advance.

A phase of American history had come to an end. Prior to 1890 the frontier had been moving continuously westwards, and Americans had to constantly adapt their institutions to a new environment. At each advance those on the frontier returned to primitive conditions and went again through all the phases of social evolution; as the frontier moved ever westwards so the people became ever more American. Turner went on to trace the movement of the frontier westwards, each major stage marked by a succession of physical features: the fall-line, the Allegheny Mountains, the Mississippi, the Missouri, the arid lands on the 99th meridian, and finally the Rockies. At each frontier the settlers had to resolve the same questions: the Indian question, the disposition of the public domain, the problem of communications with the older settlements, the extension of political organization, and the nature of religious and educational activities. Each frontier made a contribution to these problems. Turner believed that every society went through a series of evolutionary stages; the earliest was that of the hunter, then the trader, the pastoral stage of ranch life, the exploitation of the soil by unrotated crops, then the more intensive culture of denser farm settlements, and finally, the city, manufacturing and the factory system. Just as at the time he wrote, a journey from west to east would take the traveller through areas characterized by these successive economic stages, so someone standing at a point in the east would have been passed by these types in order. 'Stand at Cumberland Gap and watch the procession of civilization marching single file – the buffalo following the trail to the salt springs, the Indian, the fur trader and hunter, the cattle-raiser, the pioneer farmer – and the frontier has passed by'.[3]

As the frontier got further from the settled areas so it was necessary for the frontiersmen to produce goods with a high value per unit weight which like furs could be shipped eastwards, or like cattle could be driven eastwards. Pioneer farmers were attracted by virgin soils. There were three waves of farmers. The earliest arrivals hunted, raised cattle on the range, grew corn and vegetables; when others move into the area the frontiersman moves on to the next frontier, selling his land to new arrivals. The second wave of pioneers make roads, bridges, mills, schools and court-houses. They too then move on, and there is a final phase, the arrival of men of capital; substantial buildings are made and villages become towns; luxury replaces frugality. The continuous westward movement is prompted by the lure of free land and by the problem of soil exhaustion.

Turner then turned to the effects of the frontier upon the United States. First, he believed, it formed a melting pot in which immigrants of various nationalities became Americans. Second, it decreased the colonies' early dependence upon England. Third, it profoundly influenced federal legislation upon transport, upon tariffs, and on the disposal of the public domain. Fourth, and the most noteworthy of Turner's propositions, the frontier promoted democracy. The availability of land on the frontier gave economic equality, and this in turn led to political equality.

In his later work Turner elaborated upon these themes, rather than introducing new ones, with one important exception. In 1903 he wrote that the existence of free land in the west allowed the possibility of eastern wage earners migrating and becoming farmers, thus keeping wages high in the east and reducing labour unrest. This idea of the frontier as a safety-valve was not new, but was vigorously asserted by his followers.

Critiques of the frontier concept

Turner's view of American history appealed both to the American public and American historians, and his many students wrote widely on the themes he suggested. It was not until the 1930s that there was any substantial opposition to his ideas. Thereafter his views on the evolution of American institutions and the formation of the American character were widely attacked on a great variety of grounds. To begin with the term frontier had different meanings in his work. At some times it meant the limit of agricultural settlement, at others the whole area west of permanent settlement, at others a process which individuals on the edge of settlement experienced. Many historians have denied that the western states were noticeably democratic, and showed that the frontier states borrowed most of their political organization from the east. Others have pointed out that only a small proportion of the population was at any one time on the frontier, while others have doubted whether any explanation of American history which ignored so many other aspects of American life could be valid: '...in the single field of economics', wrote G. W. Pierson, 'he slighted the industrial revolution, he didn't seem to understand the commercial revolution, and he said nothing at all about the agricultural revolution'.[4] There is no need to pursue these aspects of Turner's theory. Our concern is simply to review those parts of his work that dealt with agrarian change, which in recent years have been profitably reinvestigated by later writers.

Types of frontier

Turner saw the westward movement as the passage of a series of economic stages; first the hunter, then the trader, then the farmer and finally the city and manufacturing. This idea was found among many nineteenth-century European historians, who believed that any society went through a sequence of economic stages. Turner's ideas seem to have been derived from the German economist List. Later he argued that the change from one type of farming to another was prompted by increasing population density. Certainly his types increase in intensity from west to east, just as in any one place on the frontier the intensity of exploitation increases as the land becomes more densely settled. This idea was of course far from new. The idea of increasing extensiveness with distance from the old settled areas – or markets – is to be found in von Thünen's work, and many of von Thünen's followers classified farming systems by increasing intensity and believed that it was caused by rising population density.

It has been argued earlier that the United States illustrates the outward progression of ever more intensive farming systems, but in the context of von Thünen's theory (see pp. 146–9). Increases in demand extended not only the limit of cultivation but pushed each zone of cultivation outwards, so that ranching was displaced westward by wheat production, and the latter by corn production. The mechanism causing increasing intensity however was rising land values, which required more intensive cultivation if profits were to be maximized (see pp. 135–40). Turner did not consider the causes of westward expansion in any detail. Among the pioneer farmers he believed there was a restlessness and a reluctance to be near other people which drove the frontiersman on, a sort of internal manifest destiny, of which the epitome was Daniel Boone. More important was the attraction that free land exercised upon those in the east, both Americans and European immigrants. The westward movement was also prompted by soil exhaustion. Pioneer farmers did not practise methods which would maintain soil fertility, and after some years of continuous cropping, would move onto virgin land, leaving newcomers to practise more intensive methods. This however was more common in the south, where cotton planters did proceed in this manner, than in the north, which was Turner's prime concern. Modern writers would probably emphasize rising demand in the east and in Western Europe as the main cause of westward expansion. A further cause of high mobility on the frontier was land appreciation. A frontier farmer could buy land cheaply beyond the edge of settlement; as the area filled up so land values rose and the pioneer could sell out at a profit and move on to repeat the sequence on the new frontier.[5]

But although on the continental scale there was an orderly progression in the occupation of unsettled land, it was by no means true on the micro-scale. In the piedmont of the Carolinas farmers and graziers moved in together. In the Great Valley there were no ranchers to precede the farmer, for land was bought from speculators and none could afford the large holdings necessary for ranching. In many regions fur trader, farmer and rancher arrived at much the same time, while in other regions some types were absent. In the arid south west the Spanish missionary played the role of the fur trader while in northern Michigan the fur trader was followed by the lumber man, not the rancher; and when the timber was exhausted it was tourism not farming that provided a living. Nor was the town the last of the economic stages to occur, as the frontier swept on and population filled in the empty spaces. Often towns were established on the frontier. St Louis is but one example.[6]

Free land, speculators and the Homestead Act

It was free land, according to Turner, that attracted settlers west; yet after 1780 the public domain was not free, but was sold by the federal government. Public land was admittedly not expensive, averaging $4.9 a hectare in 1800; it had fallen to $3.1 after the Pre-emption Act of 1841, and to 29 cents by 1854. But much of this public land was bought in advance of the line of settlement by speculators who held on and sold out at higher prices when the first farmers arrived. But it was the Homestead Act of 1862 that created the myth of free land; this allowed the pioneer 64 ha on the payment of $10, the deed being confirmed five years later, provided some improvements had taken place. But in fact the Homestead Act accounted for a small part of the public domain disposed of beyond the Mississippi. Between 1862 and 1900 19.4 million ha were settled by homesteaders. But far greater areas were sold. Railroad companies were granted 50 million ha, the federal government sold 40 million ha and much of the public domain allocated to states as a source of revenue was sold: some 56 million ha. Ironically the number of claims made under the Homestead Act between 1862 and 1890, in which year Turner believed the frontier closed, was only a quarter of those made between 1890 and 1960.[7]

Equality on the frontier

Turner believed that political democracy flowed from economic equality. Where any man could claim 64 ha, he had a sufficiency and was beholden to none. From this came both the individualism and the democracy of the frontier. Turner's vision of the small independent family farm as the basis of the American way of life was strikingly like that many have held of the English yeoman. Both were more vision than fact. In Iowa in 1860 23 per cent of the agricultural population were farm labourers, and 19 per cent in Kansas. More to the point, in Illinois in 1880 31 per cent of all farms were run by tenants, not by their owners. By 1900 35 per cent of all farms in the United States were operated by tenants, 43 per cent of all those who received income from agriculture were labourers, and only 22 per cent of the agricultural population were owner-operators. True, the national figures included the south, but even in the mid west 'free land' had failed to provide every migrant with a farm. This was partly because many who came west had too little cash to stock a farm, and worked as labourers; and because land speculators built up large estates which they leased to tenants. But even before the Homestead Act there had often been a marked inequality on the frontier. In the Shenandoah valley in the 1770s there were few landless and a rough equality in holdings. By 1800 half the houses were without land.[8]

Innovation and self-sufficiency

Turner thought the frontier farmers were essentially self-sufficient; they were too remote from the densely settled areas to market commodities, or to purchase more than the most important essentials. There is no doubt some truth in the observation; however the earliest frontier, the Atlantic frontier in Virginia, was for long dependent upon England for many essentials. As the frontier pressed westwards so connections with the east improved, and self-sufficiency broke down. Thus the first settlers of the Shenandoah valley in Virginia in the early eighteenth century, were largely self-sufficient; but they were soon driving cattle to the coast, and by the 1760s hemp was a major cash crop, later replaced by wheat, mainly for sale to the tidewater areas. By the time the frontier was crossing the Mississippi the railway provided cheap access to the east coast, so that early settlers in the Red River valley produced wheat for export east rather than attempting to provide all their own requirements. Indeed it was improved transport that broke down self-sufficiency and subsistence agriculture, not only on the frontier, but elsewhere in the United States. In 1820 it has been estimated that 75 per cent of farm produce was consumed on the farm, by 1870 45 per cent was still not sold off.[9]

Turner believed that the frontier produced not only individuality, but innovativeness. As the farmers pressed westwards they were cut off from normal supplies and had to make do with local materials. Furthermore they faced new physical environments to which they had to adjust. This was most noticeable when the arid lands of the Great Plains were first settled. Successful settlement required a variety of new techniques in farming, from the use of barbed wire to fallowing to conserve moisture. But there was a further reason for innovating. On the frontier there was a shortage of labour and wages were high. To farm an area large enough to give a good living, farmers needed labour-saving machinery such as the reaper, the gang plough, the elevator and the combine harvester. But later writers have denied the inventiveness of the frontier and shown that the bulk of the inventions were made in the east; it was however the demand from the west that prompted these inventions.[10]

The safety-valve

Turner observed that the existence of the free land on the frontier provided a safety-valve to eastern labour. When wages were low in the east labourers could become farmers in the west; this kept wages high in the east, prevented labour unrest, and delayed the development of trade unions. These ideas were by no means first put forward by Turner, but his apparent acceptance of them led many historians to incorporate the idea in their discussions of American economic and political development. It is Turner too who has had to bear the brunt of the many criticisms of the idea.

First, it has been shown that migration westwards was greatest not in periods of depression and low wages, but in periods of prosperity and rising wages. It would thus seem to have been the prospect of greater opportunities in the west rather than the oppression of low wages that prompted the movement of migrants.[11]

Second, it is difficult to show that labourers from eastern industries moved to become farmers in the west. Not only is the evidence lacking, but it would have been difficult for them to do so, on two counts. First, they had no experience in farming; and second, the cost of migration was too great. It has been seen already that land on the frontier was not free. In addition there was the cost of westward movement; the cost of buying implements, stock and breaking in virgin land; and the cost of living for a year before the first harvest. One authority has put the cost of settling in the 1850s at between $500 and $1000; as an eastern labourer earned only $300 a year, with little left to save, it is not surprising that few urban labourers left. Third, it has been argued that even if valve in the east. Further, it has been observed that although it is factories by European migrants.[12]

These attacks have prompted defenders of the safety-valve thesis to shift their ground. Accepting that few labourers from the east became farmers in the west, it is nonetheless argued that labourers and craftsmen and those in professions such as the law, teaching and banking did move west, and thus acted as a safety-valve in the east. Further, it has been observed that although it is true that there was considerable immigration from Europe, the growth of the American economy and hence of the demand for labour was sufficiently rapid for the economy to absorb this immigration and for westward migration to continue to act as a safety-valve. A further variant of the safety-valve thesis is that it was eastern farmers, not labourers, who migrated west. As their only alternative was the towns of the east, their westward movement prevented a potential labour surplus in the towns, again contributing to the relatively high real wages in the eastern United States in the nineteenth century. But to this it has been retorted that the west was quite incapable of absorbing the surplus farm population of the east. The great majority of the increased population of the United States between 1860 and 1900 went not to the western farmlands but to eastern cities, and it was the cities that provided the safety-valve.[13]

Comparative frontiers

Many other parts of the world experienced the movement of European peoples into comparatively sparsely settled areas, and so it was not surprising that historians compared the progress of agricultural settlement in other countries with Turner's views of the American frontier. The literature has however emphasized the differences between the United States and other areas rather than the similarities.

Not the least of these differences is that in environment; there is a thinly veiled streak of geographical determinism in Turner's writing, particularly in his later work on American 'sections' or regions. But the environments occupied by Europeans in the nineteenth century did differ greatly. In the United States only the Alleghenies presented any considerable barrier to westward settlement, and beyond that there was an abundance of good agricultural land until the semi-arid zone of the Great Plains was reached; the Mississippi and its tributaries provided cheap movement within the nineteenth-century mid west, and the construction of the Erie canal linked the north east coast with the interior in 1825. But northwards the initial settlements of the French in the St Lawrence and of the British and American loyalists in Ontario were hemmed in on the west by the southward extension of the Canadian shield. Not surprisingly many Canadians migrated into the United States in search of cheap land; similarly the Canadian prairies were not settled until most of the good land in the United States was occupied. American settlers formed a considerable part of the farmers who moved into the prairie provinces in the 1890s and 1900s.[14]

In Australia the mild humid coastlands of the south east were backed by uplands, which however were soon penetrated by the squatters in search of grazing land. But beyond the uplands there was no continent of good agricultural land as there was in the United States; the arid interior was reached only too soon. In Russia the occupation of the steppe lands began in the eighteenth century, and here land as fine as that to be found in the United States was settled. The peculiarity of Russia was that the good farmland stretched eastwards in a narrow zone; and as in Australia there was no river to play a role comparable with that of the Mississippi or the St Lawrence in North America. The Volga flowed south into the inland Caspian Sea, while in much of Siberia the only rivers flowed north into the Arctic. Only in South America was there a region comparable to the United States beyond the Appalachians, and that was the plains stretching north and south of the Rio de la Plata. But even here the Parana and Uruguay rivers played a limited role in agricultural advance which was primarily southwards; the railway network built by the British was far more significant.[15]

A second major difference between the United States and the other frontiers of settlement was the comparative nearness of Western Europe, in contrast to Argentina and Uruguay, South Africa and *a fortiori* to New Zealand and Australia. When the Russian frontier was advancing southwards in the Ukraine, the Black Sea afforded comparatively cheap transport to Western

Europe where the major markets for food exports were to be found; but once the Russian frontier advanced eastwards beyond the Don, the cost of movement was high.[16]

Turner saw the American frontier not only as unique, but regarded its development as a part of only American history. Yet the advance of the American frontier was only a part, although a major part, of the expansion of Western Europe generated by rapid population growth and by industrialization. It was the demand for food imports and the need for raw materials for industry that drove the frontier westwards, most obviously in the cotton belt, but also in the north. Indeed the engine of expansion was not only the demand for raw materials; British capital played a major role in the westward expansion in the United States, for the revolution of 1776 did not cut economic ties between the two countries. For most of the nineteenth century the United States was the major area of British migration of both capital and labour.

American frontier expansion differed in scale from the other frontiers; between 1860 and 1930 the area in cropland in the United States rose from 65 million hectares to 166 million hectares. In Canada, Argentina and Australia combined it increased from 0.4 million hectares to less than 60 million hectares. Thus only the Russian expansion is in any way comparable with an increase from 49 million hectares in 1860 to 109 million hectares in 1930. Nor would the inclusion of South Africa and New Zealand fundamentally alter the order of magnitude.[17]

A prime difference between the American frontier and the other frontiers was in the way in which the public domain was allocated to farmers. Although earlier it was emphasized that there was comparatively little *free* land in the United States, it was nonetheless cheaper than in the eastern United States or in Western Europe, and families of moderate means could settle on holdings large enough to provide a good living. But in much of Latin America land was held in very large holdings at a very early date. As the Argentinians advanced south in the late nineteenth century land obtained from the Indians was allocated in very large holdings. The wheat belt that developed in the very late nineteenth century was operated by Italian share-croppers, on very short leases, who did not become a class of independent farmers. In Australia the early discovery of the interior was spearheaded by graziers seeking pasture for their sheep. In spite of an attempt by the authorities to limit settlement beyond the original counties in

New South Wales in 1829, the graziers occupied much of the interior. Their tenure was made secure by an Order in Council in 1847. After the end of the gold diggings in Victoria and New South Wales, both states introduced legislation comparable to the Homestead Act in the United States. Anyone could select land in the unsettled Crown lands, including that land occupied by graziers, for £2.47 a hectare. But the graziers beat off this attack by dummying, where claims were made on behalf of the grazier by proxies, or by 'peacocking' where only the better land, particularly that with water, was bought. Thus when agricultural settlement did begin in the 1890s, it had to leap-frog the more humid areas controlled by the squatters into the semi-arid zone.[18]

Turner believed that the frontier bred democracy and individualism. But a comparison of the various frontiers suggests that life on the frontier is not a sufficient condition for the rise of democracy. The basic settlement of central and southern America was made by Spaniards and Portuguese whose background was the absolutist state of sixteenth-century Iberia; they carried these institutions to the New World with them. New England in contrast was settled by Englishmen who not only dissented from the established Church in England, but carried with them the radical ideas of the generation that overthrew Charles I. Some have suggested that the Australian frontier led to the independent nature of Australians, and their rejection of authority. But as has been noted already the nature of land tenure in Australia precluded the growth of an independent class of family farmers, and from the middle of the nineteenth century Australia was a highly urbanized society. The political ideas of nineteenth-century Australia came from radical England, not from life on the frontier.[19]

Indeed a world view of frontiers suggests that they bred servility rather than freedom. Everywhere on the frontier there was a shortage of labour, and a variety of devices were introduced to overcome this problem. In the southern colonies of British America it was slavery; in much of Spanish America the control of the labour of the agricultural Indian communities was ensured by the *repartimento* system; in Australia the early graziers relied upon convict labour; in Quebec the seigniorial system of France, although relaxed, nonetheless existed.

It is perhaps in terms of technology and land use that the similarities between the frontiers were most striking. All were geared to the growth of the West European economy, with its rising demand for grain, and later as incomes rose, meat and dairy products, so that as settlement expanded from the coastal regions, pastoral products such as wool and hides characterized the frontier, and were displaced later by more intensive products such as maize, wheat and dairy products. The comparative study of the economic geography of the nineteenth-century frontiers has yet to be undertaken. It could well be the most fruitful of the frontier studies that Turner's fertile essay has promoted.

16 Conclusions

There are nine and sixty ways of constructing tribal lays,
And-every-single-one-of-them-is-right.

RUDYARD KIPLING

It would be pleasing to end this review of the models which have been used to explain agricultural change by ranking them in order of usefulness. This cannot be done. It is unlikely that there is one key which explains all change at all places at all times. If there were it would imply that there is one single economic or cultural engine of change which determines historic change. This is unlikely although it has not prevented enthusiastic advocacy of population growth, or the class struggle or climatic change. Fashions in historical explanation change. For long Malthusian interpretations have been prevalent. In the past few years there have been increasing attempts to apply Ester Boserup's ideas to agricultural change. Geographical determinism, long dead, has been revived both in the form of climatic determinism and in the guise of eco-system studies. Who can tell which long forgotten paradigm will rise Phoenix-like from the library shelves of Academe?

Some observations can however be made. In the last thousand years in Europe a prime influence upon the nature of rural society has been the location and growth of towns. As long as the bulk of the population lived in the countryside local self-sufficiency pre-dominated, there was a lack of commercial markets for farm produce, and the high cost of transport precluded much of farm produce moving beyond the immediate locale. But since large towns became established, and particularly groups of towns, such as in northern Italy, Flanders or Holland, there arose a demand for farm produce: between the town and the surrounding countryside there arose a complex interaction which gave rise to more intensive farming, higher land values, the purchase of land by urban merchants, easier access to credit, the easier diffusion of new ideas, and markets for products used as raw materials in industry. But with increasing distance from the urban markets, farming became less commercialized, less intensive, less innovating.

Within Europe the long term rise and fall of population – from 1000 to 1340, and from 1450 to the 1650s – provided both a stimulus to change and a cause of expansion. In some areas higher population densities have had adverse consequences upon agricultural society; in others they have led to progressive intensification and improvements of farming methods. But the effects of population growth were most dramatic in the eighteenth and nineteenth centuries. Not only did population grow rapidly, but the numbers and the proportion living in towns increased dramatically, providing a large market for the farmer; the fall in transport cost extended the area influenced by the cities, until ultimately New Zealand and Patagonia were linked to London and Philadelphia. But it was not now merely greater numbers that prompted change, but greater wealth. The industrial revolution provided eventually a steady increase in income, and changed the demand for agricultural products from cereals to meat and milk, while manufacturing industries sought more wool, more cotton, more vegetable oils and more rubber and silk. New industries rose to aid the farmer in producing new goods: the interaction between farmer and town intensified.

Such a view of European rural history leaves much unsaid. The great variations in climate and land tenure ensured that the reactions of different regions to the prompting of first population changes and then income changes would not be the same. It is hard to see climate as a prime mover; but it would be a major cause of differing reactions to external stimuli, be it price or population pressure. Languedoc and Lapland are poles apart in climate as well as economy. On a smaller scale the east and west of Britain differed not only in climate but in social and farm structure. The rise of capitalist agriculture in eastern England was possible because of the long established tenurial and social conditions. The very different conditions of the west precluded such a development; it was not merely a matter of slope and soil or rain and run-off.

In short a major conclusion of this book must be that to search for a prime mover to explain agrarian change is to pursue a chimera. The same force will work its way out in different ways in different regions. Indeed perhaps it can be argued that the way ahead lies in the application of specific models not simply to one area, but to several regions. Comparative studies may give the agricultural historian comparative advantage.

Notes and references

Chapter 2 The adverse consequences of population growth

1 R. Bićanić, 'Three concepts of agricultural overpopulation', in R. N. Dixey (ed.), *International Explorations of Agricultural Economics* (Ames, Iowa, 1964), pp. 8–20; C. J. Robertson, 'Population and agriculture with special reference to agricultural overpopulation', in International Institute of Agriculture, *Documentation for the European Conference on Rural Life, 1939* (Rome, 1939), pp. 11–30.

2 C. Clark, *Population Growth and Land Use* (London, 1969); E. Boserup, *The Conditions of Agricultural Growth: The Economics of Agrarian Change under Population Pressure* (London, 1965).

3 T. R. Malthus, *An Essay on the Principle of Population or a View of its Past and Present Effects on Human Happiness*, ed. G. T. Bettany (London, 1890).

4 A. T. Peacock, 'Economic theory and the concept of an optimum', in J. B. Craggs and N. W. Pirie (eds.), *The Numbers of Man and Animals*, Institute of Biology (London, 1955), pp. 1–7; J. N. Sinha, 'Population and agriculture', in L. Tabah (ed.), *Population Growth and Economic Development in the Third World* (Dolhain, Belgium, 1976), vol. 2, pp. 251–306.

5 J. D. Stryker, 'Optimum population in rural areas: empirical evidence from the Franc zone', *Quarterly Journal of Economics*, vol. 91 (1977), pp. 177–93.

6 B. Kenadjiian, 'Disguised unemployment in underdeveloped countries', *Zeitschrift für Nationalökonomie*, vol. 21 (1961), pp. 216–23; W. C. Robinson, 'Types of disguised rural underemployment and some policy implications', *Oxford Economic Papers*, vol. 21 (1969), pp. 373–86.

7 W. E. Moore, *The Economic Demography of Eastern and Southern Europe* (Geneva, 1945).

8 P. Sen Gupta, 'Population and resource development in India', in W. Zelinsky, L. Kosinski and R. Mansell Prothero (eds.), *Geography and a Crowding World* (Oxford, 1970), pp. 424–41.

9 P. N. Rosenstein-Rodan, 'Disguised unemployment and underemployment in agriculture', *Monthly Bulletin of Agricultural Economics and Statistics*, vol. 6 (1957), pp. 1–7.

10   W. Allan, *The African Husbandman* (Edinburgh, 1965).

11   R. L. Carneiro, 'From autonomous villages to the state, a numerical estimate', in B. Spooner (ed.) *Population Growth: Anthropological Implications* (Boston, Mass., 1972), pp. 64–74; H. C. Brookfield and P. Brown, *Struggle for Land: Agriculture and Group Territories among the Chimbu of the New Guinea Highlands* (Melbourne, 1963), pp. 105–24; J. M. Street, 'An evaluation of the concept of carrying capacity', *Professional Geographer*, vol. 21 (1969), pp. 104–7; T. Bayliss-Smith, 'Maximum populations and standard populations: the carrying capacity question', in D. Green, C. Hazelgrove and M. Spriggs (eds.), *Social Organisation and Settlement; Contributions from Anthropology, Archaeology and Geography*, British Archaeological Reports no. 47 (Oxford, 1978), pp. 129–34; R. Feachem, 'A classification of carrying capacity formulae', *Australian Geographical Studies*, vol. 11 (1973), pp. 234–6.

12   D. B. Grigg, 'Population pressure and agricultural change', *Progress in Geography*, vol. 8 (1976), pp. 135–76.

13   R. Minami, 'An analysis of Mathus' population theory', *Journal of Economic Behaviour*, vol. 1 (1961), pp. 53–63.

14   R. Cameron, 'The logistics of European economic growth: a note on historical periodization', *Journal of European Economic History*, vol. 2 (1973), pp. 145–8.

15   W. Abel, *Agricultural Fluctuations in Europe: From the Thirteenth to the Twentieth Centuries* (London, 1980); E. Le Roy Ladurie, *The Peasants of Languedoc* (Urbana, Ill., 1974); B. H. Slicher van Bath, *The Agrarian History of Western Europe, AD 500–1850* (London, 1963); Jan de Vries, *The Dutch Rural Economy in the Golden Age, 1500–1700* (New Haven, 1974).

16   M. M. Postan, 'Agrarian society in its prime: England', in M. M. Postan (ed.), *Cambridge Economic History of Europe:* vol. 1, *The Agrarian Life of the Middle Ages* (Cambridge, 1966), pp. 548–632.

17   Postan, 'Agrarian society'; J. Z. Titow, *English Rural Society 1200–1350* (London, 1969).

18   J. C. Russell, 'The pre-plague population of England', *Journal of British Studies*, vol. 5 (1966), pp. 1–21; R. H. Hilton, 'Rent and capital formation in feudal society', *Second Conference of Economic History, Aix en Provence: Congrès et colloques* (Paris, 1965), vol. 8, pp. 33–68.

19   K. H. Connell, *The Population of Ireland, 1750–1845* (Oxford, 1950).

20   P. M. A. Bourke, 'The use of the potato in pre-famine Ireland', *Journal of the Statistical and Social Inquiry Society of Ireland*, vol. 21 (1967–8), pp. 72–96; P. M. A. Bourke, 'The agricultural statistics of the 1841 Census of Ireland: a critical review', *Economic History Review*, vol. 18 (1965), pp. 376–91; R. O. Crotty, *Irish Agricultural Production: Its Volume and Structure* (Cork, 1966).

Chapter 3   The positive consequences of population growth

1   J. Nõu, *Studies in the Development of Agricultural Economics in Europe* (Uppsala, 1967).
2   E. Boserup, *The Conditions of Agricultural Growth; The Economics of Agrarian Change under Population Pressure* (London, 1965); E. Boserup, *Population and Technology* (Oxford, 1981).
3   E. J. Nell, 'The technology of intimidation', *Peasant Studies Newsletter*, vol. 1 (1972) pp. 39–44; P. E. L. Smith, 'Land use, settlement patterns and subsistence agriculture: a demographic perspective', in P. J. Ucko, R. Tringham and G. Dimbleby (eds.), *Man, Settlement and Urbanism* (London, 1972), pp. 409–25; D. Grigg, 'Ester Boserup's theory of agrarian change; a critical review', *Progress in Human Geography*, vol. 3 (1975) pp. 64–84; J. de Vries, *The Dutch Rural Economy in the Golden Age, 1500–1700* (New Haven, Conn., 1974).
4   Boserup, *Population and Technology*, pp. 23–4, 59.
5   Boserup, *The Conditions of Agricultural Growth*, p. 117.
6   de Vries, *The Dutch Rural Economy*, pp. 68, 125, 137–55; H. van der Wee, *The Growth of the Antwerp Market and the European Economy* (The Hague, 1962), vol. 2, pp. 113, 116, 122, 169, 209.
7   C. Vandenbroeke, 'Cultivation and consumption of the potato in the seventeenth and eighteenth centuries', *Acta Historiae Neerlandica*, vol. 5, pp. 15–39; M. Drake, *Population and Society in Norway 1735–1865* (Cambridge, 1969); L. M. Cullen, 'Irish history without the potato', *Past and Present*, vol. 40 (1968) pp. 72–83; M. Morineau, 'La pomme de terre au XVIIIe siècle', *Annales ESC*, vol. 25 (1970), pp. 1767–85.

Chapter 4   Price, technology and environment

1   E. Huntington, *The Pulse of Asia* (Boston, Mass., 1907); E. Huntington, *Mainsprings of Civilisation* (New York, 1945); H. T. Buckle, *History of Civilisation in England* (London, 1903), 3 vols.; A. J. Toynbee, *A Study of History* (London, 1939–61), 12 vols.; K. Marx, *Capital* (London, 1977), p. 481.
2   L. Febvre, *A Geographical Introduction to History* (London, 1950); G. Tatham, 'Environmentalism and possibilism', in G. Taylor (ed.), *Geography in the Twentieth Century: A Study of Growth, Fields, Techniques, Aims and Trends* (London, 1953), pp. 128–62.
3   E. Huntington, 'Climatic change and agricultural exhaustion as elements in the fall of Rome', *Quarterly Journal of Economics*, vol. 31 (1917), pp. 173–208.
4   J. P. Bakker, 'The significance of physical geography and pedology for historical geography in the Netherlands', *Tijdschrift voor*

Economische en Sociale Geographie, vol. 49 (1958), pp. 214–20.

5   W. L. Thomas, Jr (ed.), *Man's Role in Changing the Face of the Earth* (Chicago, 1956).

6   L. P. Smith, 'The effect of weather, drainage efficiency and duration of spring cultivations on barley yields in England', *Outlook on Agriculture*, vol. 7 (1972), pp. 79–84.

7   D. K. Britton, *Cereals in the United Kingdom: Production, Marketing, Utilisation* (Oxford, 1969), pp. 94–5, 695–70; L. P. Smith, 'Meteorology and the pattern of British grassland farming', *Agricultural Meteorology*, vol. 4 (1967), pp. 321–35.

8   D. Ricardo, *The Principles of Political Economy and Taxation* (London, 1957), pp. 33–45.

9   K. H. W. Klages, *Ecological Crop Geography* (New York, 1942).

10 D. Grigg, *The Agricultural Revolution in South Lincolnshire* (Cambridge, 1966), p. 38.

11 B. H. Slicher van Bath, *The Agrarian History of Western Europe, AD 500–1850* (London, 1963), pp. 114–15.

12 E. H. Phelps Brown and S. V. Hopkins, 'Seven centuries of the price of consumables, compared with builders' wage-rates', *Economica*, vol. 23 (1956), pp. 296–314.

13 Slicher van Bath, *The Agrarian History*, pp. 100, 105, 112–231.

14 ibid., p. 133; M. L. Parry, *Climatic Change, Agriculture and Settlement* (Folkestone, 1975); M. M. Postan, 'Medieval agrarian society in its prime: England', in M. M. Postan (ed.), *The Cambridge Economic History of Europe*; vol. 1, *The Agrarian Life of the Middle Ages* (Cambridge, 1966), pp. 548–52.

15 M. Williams, 'The enclosure and reclamation of wasteland in England and Wales in the eighteenth and nineteenth centuries', *Transactions and Papers of the Institute of British Geographers*, vol. 51 (1970), pp. 55–70.

16 A. R. H. Baker, 'Evidence in the *Nonarum Inquisitiones* of the contracting arable lands in England during the early fourteenth century', *Economic History Review*, vol. 19 (1966), pp. 518–32.

17 M. W. Beresford, 'A review of historical research (to 1968)', in M. Beresford and J. G. Hurst (eds.) *Deserted Medieval Villages: Studies* (London, 1971), pp. 3–75.

18 A. John, 'Agricultural productivity and economic growth in England, 1700–1760', *Journal of Economic History*, vol. 25 (1965), pp. 19–34; E. L. Jones, 'Agriculture and economic growth, 1660–1750: agricultural change', *Journal of Economic History*, vol. 25 (1965), pp. 1–18.

19 J. T. Coppock, 'The changing face of England: 1850–*circa* 1900', in H. C. Darby (ed.), *New Historical Geography of England* (Cambridge, 1973), pp. 607–12; P. Perry, *British Farming in the Great Depression 1870–1914: An Historical Geography* (Newton Abbot, 1974), p. 75.

20 Grigg, *The Agricultural Revolution*, pp. 137–41; E. Dunsdorfs, *The*

Australian Wheat Growing Industry 1788–1948 (Melbourne, 1956), pp. 154–9.

21 A. H. John, 'The course of agricultural change, 1660–1760', in L. S. Pressnell (ed.), *Studies in the Industrial Revolution Presented to T. S. Ashton* (London, 1960), pp. 129–34; D. Grigg, 'Changing regional values during the agricultural revolution in South Lincolnshire', *Transactions and Papers of the Institute of British Geographers*, vol. 30 (1962), pp. 91–105; H. C. Darby, 'The draining of the English claylands', *Geographische Zeitschrift*, vol. 52 (1964), pp. 190–201; E. J. T. Collins and E. L. Jones, 'Sectoral advance in English agriculture 1850–1880', *Agricultural History Review*, vol. 15 (1967), pp. 65–81; R. W. Sturgess, 'The agricultural revolution on the English clays', *Agricultural History Review*, vol. 14 (1966), pp. 104–21; E. H. Whetham, 'Sectoral advance in English agriculture: a summary', *Agricultural History Review*, vol. 16 (1968), pp. 46–8; Jones, 'Agriculture and economic growth'; John, 'Agricultural productivity and economic growth', pp. 1–18; A. D. M. Phillips, 'Underdraining and the English claylands, 1850–1880: a review', *Agricultural History Review*, vol. 17 (1969), pp. 44–55.

22 E. J. Russell, *Soil Conditions and Plant Growth* (London, 1956), pp. 595–9; G. V. Jacks, *Soil* (London, 1954), pp. 21–3; Agricultural Advisory Council, *Modern Farming and the Soil* (London, 1970), pp. 20, 25, 29; L. P. Smith, *The Significance of Winter Rainfall over Farmland in England and Wales*, Ministry of Agriculture, Fisheries and Food, *Technical Bulletin 24* (London, 1971).

23 Grigg, 'Changing regional values'; Grigg, *The Agricultural Revolution*.

24 D. Grigg, *The Agricultural Systems of the World: An Evolutionary Approach* (Cambridge, 1974), p. 277.

25 W. A. Mackintosh, *Prairie Settlement; The Geographical Setting* (Toronto, 1934), pp. 17, 19.

26 ibid., pp. 79, 170; C. T. Wilson, *A Century of Canadian Grain: Government Policy to 1951* (Saskatoon, 1978), p. 10; J. W. Morrison, 'Marquis wheat – a triumph of scientific endeavour', *Agricultural History*, vol. 34 (1960), pp. 182–8.

27 W. A. Douglas Jackson, 'The Russian non-chernozem wheat base', *Annals of the Association of American Geographers*, vol. 49 (1959), pp. 97–109; B. Fullerton, 'The northern margin of grain production in Sweden in the twentieth century', *Transactions and Papers of the Institute of British Geographers*, vol. 20 (1954), pp. 181–91; R. R. Platt (ed.), *Finland and its Geography* (London, 1957), p. 123; U. Varjo, 'Productivity and fluctuating limits of crop cultivation in Finland', *Geographica Polonica*, vol. 40 (1979), pp. 225–33.

Chapter 5 Agricultural systems as ecosystems

1 A. G. Tansley, 'The use and abuse of vegetational concepts and terms', *Ecology*, vol. 16 (1935), pp. 284–307.

2 R. W. Snaydon and J. Elston, 'Flows, cycles and yields in agricultural ecosystems', in A. N. Duckham, J. G. W. Jones and E. H. Roberts, (eds.), *Food Production and Consumption: The Efficiency of Human Food Chains and Nutrient Cycles* (Amsterdam, 1976), p. 19.

3 M. J. Frissel (ed.), *Cycling of Mineral Nutrients in Agricultural Ecosystems* (Oxford, 1978), p. 16.

4 A. N. Duckham and G. Masefield, *Farming Systems of the World* (London, 1970), p. 10.

5 K. H. Connell, 'The potato in Ireland', *Past and Present*, vol. 23 (1962), pp. 57–63; C. Vandenbroeke, 'Cultivation and consumption of the potato in the seventeenth and eighteenth centuries', *Acta Historia Neerlandica*, vol. 5 (1971), pp. 15–39; M. Drake, *Population and Society in Norway, 1735–1865* (London, 1969), pp. 56–63.

6 M. P. Miracle, 'The introduction and spread of maize in Africa', *Journal of African History*, vol. 6 (1966), pp. 39–55; W. O. Jones, 'Manioc: an example of innovation in African economies', *Economic Development and Cultural Change*, vol. 5 (1957), pp. 100–10; P. Ho, 'The introduction of American food plants into China', *American Anthropologist*, vol. 57 (1955), pp. 191–201.

7 E. J. Russell, *Soil Conditions and Plant Growth* (London, 1961), p. 318.

8 W. Linnard, 'Terms and techniques in shifting cultivation in Russia', *Tools and Tillage*, vol. 1 (1970), pp. 192–7; S. Montelius, 'The burning of forest land for the cultivation of crops', *Geografiska Annaler*, vol. 35 (1953), pp. 41–54.

9 A. P. Usher, 'Soil fertility, soil exhaustion and their historical significance', *Quarterly Journal of Economics*, vol. 37 (1923), pp. 385–411; R. Lennard, 'The alleged exhaustion of the soil in medieval England', *Economic Journal*, vol. 32 (1922), pp. 12–27; W. S. Cooter, 'Ecological dimensions of medieval agrarian systems', *Agricultural History*, vol. 52 (1978), pp. 458–77.

10 G. P. H. Chorley, 'The agricultural revolution in northern Europe, 1750–1880; nitrogen, legumes and crop productivity', *Economic History Review*, vol. 34 (1981), pp. 71–93.

11 R. S. Loomis, 'Ecological dimensions of medieval agrarian systems: an ecologist responds', *Agricultural History*, vol. 52 (1978), pp. 478–83.

12 Chorley, 'The agricultural revolution', pp. 71–93.

13 Frissel (ed.), *Cycling of Mineral Nutrients*, pp. 19–21; D. L. Farmer, 'Grain yields on the Winchester manors in the later Middle Ages',

Economic History Review, vol. 30 (1977), pp. 555–66; H. V. Garner and G. V. Dyke, 'The Broadbalk Yields', *Rothamsted Experimental Station Report for 1968* (1969), pp. 26–49.

14 E. Kerridge, *The Agricultural Revolution* (London, 1967).

15 Chorley, 'The agricultural revolution', pp. 71–93; Kerridge, *The Agricultural Revolution*.

16 G. Borgstrom, *The Food and People Dilemma* (North Scituate, Mass., 1973), pp. 46–7.

17 T. Jacobsen and R. M. Adams, 'Salt and silt in ancient Mesopotamian agriculture', *Science*, vol. 128 (1958), pp. 1251–8.

Chapter 6   Climatic change and farming history

1 E. Le Roy Ladurie, *Times of Feast, Times of Famine: A History of Climate Since the Year 1000* (London, 1972); E. Le Roy Ladurie and M. Baulant, 'Grape harvests from the fifteenth through the nineteenth centuries', *Journal of Interdisciplinary History*, vol. 10 (1980) pp. 839–945; J. A. Kington, 'An application of phenological data to historical climatology', *Weather*, vol. 29 (1974), pp. 320–8; J. de Vries, 'Histoire du climat et économie; des faits nouveaux, une interprétation différente', *Annales ESC*, vol. 32 (1977), pp. 198–226.

2 ibid.; W. Dansgard, S. J. Johnsen, N. Rech, N. Gundestrup, H. B. Clausen and C. U. Hammer, 'Climatic changes, Norsemen and modern man', *Nature*, vol. 255 (1975), pp. 24–7; H. E. Landsberg, 'Past climates from unexploited written sources', *Journal of Interdisciplinary History*, vol. 10 (1980), pp. 631–42; F. Sandon, 'A millennium of West European climate – a cu-sum look at dendroclimatology', *Weather*, vol. 29 (1974), pp. 162–6.

3 H. H. Lamb, 'Britain's changing climate', in C. G. Johnson and L. P. Smith (eds.), *The Biological Significance of Climatic Changes in Britain*, Symposium of the Institute of Biology no. 14 (London, 1965), pp. 4–23; H. H. Lamb, *Climate: Present, Past and Future* (London, 1977), vol. 2, pp. 267–463; J. Gribbin and H. H. Lamb, 'Climatic change in historical times', in J. Gribbin (ed.), *Climatic Change* (Cambridge, 1978), pp. 68–80.

4 E. Le Roy Ladurie, 'History and climate', in P. Burke (ed.), *Economy and Society in Early Modern Europe* (London, 1972), pp. 134–69; J. de Vries, 'Measuring the impact of climate on history: the search for appropriate methodologies', *Journal of Interdisciplinary History*, vol. 10 (1980), pp. 599–630; B. H. Slicher van Bath, *The Agrarian History of Western Europe, AD 500–1850* (London, 1963), p. 7; J. L. Anderson, 'Climate and the historians', in A. B. Pittock (ed.), *Climatic Change and Variability: A Southern Perspective* (Cambridge, 1978), pp. 310–13.

5 H. Lucas, 'The great European famine of 1315, 1316 and 1317', *Speculum*, vol. 5 (1930), pp. 343–77; L. P. Smith, 'The significance of

climatic variations in Britain', in UNESCO, *Changes of Climate*, Arid Zone Research no. 20 (Paris, 1963), pp. 455–63.

6 J. Z. Titow, 'Evidence of weather in the account rolls of the bishopric of Winchester 1209–1350', *Economic History Review*, vol. 12 (1960), pp. 460–7; W. G. Hoskins, 'Harvest fluctuations and English economic history, 1620–1759', *Agricultural History Review*, vol. 16 (1968), pp. 15–31.

7 J. W. King, E. Hurst, A. J. Slater, P. A. Smith and B. Tamkin, 'Agriculture and sunspots', *Nature*, vol. 252 (1974), pp. 2–3; R. A. Bryson, 'A perspective on climatic change', *Science*, vol. 184 (1974), pp. 753–54; 'Cultural, economic and climatic records', in Pittock (ed.), *Climatic Change and Variability*, p. 324; M. L. Parry, 'The significance of the variability of summer warmth in upland Britain', *Weather*, vol. 31 (1976), pp. 212–17.

8 M. L. Parry, *Climatic Change, Agriculture and Settlement* (Folkestone, 1978), p. 65; P. Alexandre, 'Les variations climatiques au moyen âge', *Annales ESC*, vol. 32 (1977), pp. 183–95; Lamb, *Climate*, p.277.

9 Parry, *Climatic Change*, pp. 160–2; H. H. Lamb, 'Climatic change and foresight in agriculture: the possibilities of long term weather advice', *Outlook on Agriculture*, vol. 7 (1973), pp. 203–10; Lamb, 'Britain's changing climate', pp. 4–23; Lamb, *Climate*, p. 461; G. Utterström, 'Climatic fluctuations and population problems in early modern history', *Scandinavian Economic History Review*, vol. 3 (1955), pp. 3–47; G. Manley, *Climate and the British Scene* (London, 1952), p. 238; C. J. Harrison, 'Grain price analysis and harvest qualities, 1465–1634', *Agricultural History Review*, vol. 19 (1971), pp. 135–55.

10 G. Manley, 'Central England temperatures; monthly means 1659 to 1973', *Quarterly Journal of the Royal Meteorological Society*, vol. 100 (1974), pp. 389–405; S. Gregory, 'The definition of wet and dry periods for discrete regional units', *Weather*, vol. 34 (1979), pp. 363–9; M. Overton, 'Estimating crop yields from probate inventories: an example from East Anglia, 1585–1735', *Journal of Economic History*, vol. 39 (1979), pp. 363–78; A. H. John, 'English agricultural improvements and grain exports 1660–1765', in D. C. Coleman and A. H. John (eds.), *Trade, Government and Economy in Pre-industrial England* (London, 1976), p. 48.

11 M. L. Bunce, 'The agricultural depression in South Yorkshire and North Nottinghamshire, 1875–1900: climatic hazard and price competition', *Canadian Geographer*, vol. 4 (1972), pp. 323–37.

Chapter 7   The nature of peasant societies

1 P. Hall (ed.), *Von Thünen's Isolated State* (Oxford, 1966).

2 J. Nõu, *Studies in the Development of Agricultural Economics in Europe* (Uppsala, 1967).

3 K. Marx, *Pre-capitalist Economic Formations* (London, 1964).

4 A. V. Chayanov, *The Theory of Peasant Economy*, ed. D. Thorner, B. Kerblay and R. E. F. Smith (Homewood, Ill., 1966).

5 J. H. Boeke, *Economics and Economic Policy of Dual Societies* (New York, Institute of Pacific Relations, 1953).

6 Polly Hill, 'A plea for indigenous economics: the West African example', *Economic Development and Cultural Change*, vol. 15 (1966), pp. 10–20.

7 K. Polanyi, 'The economy as instituted process', in K. Polanyi, C. M. Arensberg and H. W. Pearson (eds.), *Trade and Market in the Early Empires: Economies in History and Theory* (New York, 1965); G. Dalton, 'Economic theory and peasant society', *American Anthropology*, vol. 63 (1961), pp. 1–25; G. Dalton, 'Traditional production in primitive African economies', *Quarterly Journal of Economics*, vol. 76 (1962), pp. 360–78.

8 E. R. Wolf, *Peasants* (Englewood Cliffs, NJ, 1966); T. Shanin, 'The nature and logic of the peasant economy', *Journal of Peasant Studies*, vol. 1 (1973), pp. 63–80.

9 E. L. Jones (ed.), *European Peasants and their Markets: Essays in Agrarian Economic History* (Princeton, NJ, 1975).

10 Chayanov, *The Theory of Peasant Economy*; see also M. Harrison, 'Chayanov and the economics of the Russian peasantry', *Journal of Peasant Studies*, vol. 2 (1975), pp. 389–417; J. R. Millar, 'A reformulation of A. V. Chayanov's theory of the peasant economy', *Economic Development and Cultural Change*, vol. 18 (1969), pp. 219–29.

11 In Chayanov, *The Theory of Peasant Economy*, pp. 1–28.

12 In ibid., pp. 31–269.

13 M. Nash, *Primitive and Peasant Economic Systems* (San Francisco, 1966), pp. 20–35; Wolf, *Peasants*, pp. 2–15; Shanin, 'The nature and logic', pp. 63–8.

14 P. A. Sorokin, C. C. Zimmerman and C. J. Galpin, *A Systematic Sourcebook in Rural Sociology*, vol. 2 (New York, 1931), p. 128; S. H. Cousens, 'Changes in Bulgarian agriculture', *Geography*, vol. 52 (1967), p. 12; R. Bićanić, 'The effect of the war on rural Yugoslavia', *Geographical Journal*, vol. 103 (1944), p. 39.

15 P. T. Bauer and B. S. Yamey, 'A case study of response to price in an underdeveloped country', *Economic Journal*, vol. 69 (1959), pp. 800–7; E. R. Dean, 'Economic analysis and African responses to price', *Journal of Farm Economics*, vol. 47 (1965), pp. 402–9; R. Krishna, 'Farm supply response in India–Pakistan: a case study of the Punjab region', *Economic Journal*, vol. 73 (1963), pp. 477–87; R. M. Stern, 'The price responsiveness of primary producers', *Review of Economics and Statistics*, vol. 44 (1962), pp. 202–7.

16 T. W. Schultz, *Transforming Traditional Agriculture* (New Haven,

Conn., 1963).

17   F. Dovring, 'Development as history', *Comparative Studies in Society and History*, vol. 21 (1979), pp. 557–71.

18   M. Lipton, 'Should reasonable farmers respond to price changes?', *Modern Asian Studies*, vol. 1 (1966), pp. 95–9; A. Parik, 'Market responsiveness of peasant cultivators: some evidence from pre-war India', *Journal of Development Studies*, vol. 8 (1972), pp. 289–306; A. I. Medani, 'The supply response to price of African farmers at various stages of development', *Oxford Agrarian Studies*, vol. 1 (1972), pp. 57–62; Q. Paris, 'The farmer and the norms of a market economy in developing countries; an analysis of case studies', *Farm Economist*, vol. 11 (1970), pp. 493–510.

19   A. Everitt, 'Farm labourers', in Joan Thirsk (ed.), *The Agrarian History of England and Wales*; vol. 4, *1500–1640* (Cambridge, 1967), p. 425; Chayanov, *The Theory of Peasant Economy*, p. 14.

20   Jan de Vries, *The Dutch Rural Economy in the Golden Age, 1500–1700* (New Haven, Conn., 1974), pp. 119–36.

21   B. Galeski, *Basic Concepts of Rural Sociology* (Manchester, 1972), pp. 16–17; S. von Frauendorfer, 'American farmers and European peasantry', in Sorokin, Zimmerman and Galpin, *A Systematic Sourcebook*, vol. 2, p. 166; R. Hilton, 'Medieval peasants; any lessons', *Journal of Peasant Studies*, vol. 1 (1973–4), pp. 206–19; T. C. Blegen, *Norwegian Migration to America 1825–1860* (Minneapolis, 1931), p. 5; D. S. Thomas, *Social and Economic Aspects of Swedish Population Movements, 1750–1933* (New York, 1940), p. 95.

22   R. D. Crotty, *Irish Agricultural Production: Its Volume and Structure* (Cork, 1966), p. 44; A. R. Bridbury, 'The Black Death', *Economic History Review*, vol. 26 (1973–4), p. 578.

23   M. Goldschmidt and E. J. Kunkel, 'The structure of the peasant family', *American Anthropologist*, vol. 73 (1971), pp. 1058–76; M. N. Diaz, 'Introduction: economic relations in peasant society', in J. M. Potter, M. N. Diaz and G. M. Foster (eds.), *Peasant Society: A Reader* (Boston, Mass., 1967), p. 4.

24   Wolf, *Peasants*, p. 3; G. Foster, 'Introduction: What is a Peasant?', in Potter, Diaz and Foster, *Peasant Society*, p. 4; G. Dalton, 'How exactly are peasants exploited?', *American Anthropologist*, vol. 76 (1974), pp. 553–61.

25   M. W. Flinn, 'The stabilisation of mortality in pre-industrial Europe', *Journal of European Economic History*, vol. 3 (1974), pp. 287–317; J. D. Post, 'Famine, mortality and epidemic disease in the process of modernization', *Economic History Review*, vol. 29 (1976), pp. 14–37; J. Tarrant, *Food Policies* (London, 1980).

26   M. Lipton, 'The theory of the optimising peasant', *Journal of Development Studies*, vol. 4 (1968), pp. 327–51.

Chapter 8 Structural transformation and turning points

1 D. Grigg, 'The world's agricultural labour force 1800–1970', *Geography*, vol. 60 (1975), pp. 194–202; P. Bairoch and J. Limbor, 'Changes in the industrial distribution of the world labour force by region, 1880–1960', *International Labour Review*, vol. 98 (1968), pp, 311–66; FAO, *Production Yearbook 1979*, vol. 33 (1980), p. 61.

2 Paul Bairoch, 'Agriculture and the industrial revolution 1700–1914', in C. M. Cipolla (ed.), *The Fontana Economic History of Europe*; vol. 3, *The Industrial Revolution* (London, 1973), p. 458; P. Mathias, 'The social structure in the eighteenth century: a calculation by Joseph Massie', *Economic History Review*, vol. 10 (1957), pp. 30–45; Jan de Vries, *The Dutch Rural Economy in the Golden Age 1500–1700* (New Haven, Conn., 1974), p. 87.

3 Paul Bairoch, 'Population urbaine et taille des villes en Europe de 1600 à 1970: presentation de séries statistiques', *Revue d'Histoire Economique et Sociale*, vol. 54 (1976), p. 312.

4 B. F. Johnston, 'Comment: sectoral interdependence, structural transformation and agricultural growth', in C. R. Wharton (ed.), *Subsistence Agriculture and Economic Development* (London, 1970), pp. 348–53.

5 F. Dovring, 'The share of agriculture in a growing population', *Monthly Bulletin of Agricultural Economics and Statistics*, vol. 8 (1959), pp. 1–11; Johnston, 'Comment', pp. 348–53.

6 D. W. Jorgenson, 'The role of agriculture in economic development: classical versus neoclassical models of growth', in Wharton (ed.), *Subsistence Agriculture*, p. 343.

7 C. M. Law, 'Some notes on the urban population of England and Wales in the eighteenth century', *Local Historian*, vol. 10 (1972), pp. 13–26; C. M. Law, 'The growth of the urban population in England and Wales, 1801–1911', *Transactions and Papers of the Institute of British Geographers*, vol. 41 (1967), pp. 125–44.

8 R. Bićanić, *Turning Points in Economic Development* (The Hague, 1972), pp. 155–88.

9 E. L. Jones, 'The agricultural labour market in England, 1793–1872', *Economic History Review*, vol. 17 (1964–5), pp. 322–38; E. J. Hobsbawm and G. Rudé, *Captain Swing* (London, 1969); E. J. T. Collins, *From Sickle to Combine* (Reading, 1969); E. J. T. Collins, 'Harvest technology and labour supply in Britain, 1790–1870', *Economic History Review*, vol. 22 (1969), pp. 453–73; P. A. David, *Technological Choice, Innovation and Economic Growth: Essays on American and British Experience in the Nineteenth Century* (London, 1975), pp. 214–18; E. J. T. Collins, 'Labour supply and demand in European agriculture 1800–1880', in E. L. Jones and S. J. Woolf (eds.), *Agrarian Change and Economic Development: The Historical Problems* (London,

1969), pp. 61–94.

10 Y. Hayami and V. W. Ruttan, *Agricultural Development: An International Perspective* (Baltimore, Md., 1971).

Chapter 9 Industrialization, demand and new technologies

1 FAO, *Production Yearbook 1979*, vol. 33 (1980), p. 250; FAO, *Fourth World Food Survey* (Rome, 1977), p. 53.

2 J. C. Toutain, 'La consommation alimentaire en France de 1789 à 1964', *Economies et sociétiés*, vol. 5 (1971), p. 1977; C. Lis and H. Soly, 'Food consumption in Antwerp between 1807 and 1859: a contribution to the standard of living debate', *Economic History Review*, vol. 30 (1977), pp. 460–86; W. Abel, *Agricultural Fluctuations in Europe: From the Thirteenth to the Nineteenth Centuries* (London, 1980), pp. 292–7.

3 M. de Nigris, 'The impact of urbanization on food demand', *Monthly Bulletin of Agricultural Economics and Statistics*, vol. 22 (1973), pp. 1–16; Abel, *Agricultural Fluctuations in Europe*, p. 142; P. Bairoch, 'Agriculture and the industrial revolution 1700–1914', in C. M. Cipolla (ed.), *The Fontana Economic History of Europe*; vol. 3, *The Industrial Revolution* (London, 1973), p. 469; E. M. Ojala, *Agriculture and Economic Progress* (Oxford, 1952), p. 101; E. J. T. Collins, 'The "consumer revolution" and the growth of factory foods: changing patterns of bread and cereal eating in Britain in the twentieth century', in D. Oddy and D. Miller (eds.), *The Making of the Modern British Diet* (London, 1976), p. 26.

4 E. J. T. Collins, 'Dietary change and cereal consumption in Britain in the nineteenth century', *Agricultural History Review*, vol. 23 (1975), pp. 97–115.

5 P. K. O'Brien, D. Heath and C. Keyder, 'Agriculture in Britain and France 1815–1914', *Journal of European Economic History*, vol. 6 (1977), pp. 339–92; J. H. Kirk, 'The agricultural industry, an introduction', in A. Edwards and A. Rogers (eds.), *Agricultural Resources: An Introduction to the Farming Industry of the United Kingdom* (London, 1974), p. 8.

6 E. Jensen, *Danish Agriculture, its Economic Development: A Description and Economic Analysis Centering on the Free Trade Epoch, 1876–1930* (Copenhagen, 1937); T. W. Freeman, *Ireland: A General and Regional Geography* (London, 1909), p. 200.

7 E. B. Shaw, 'Recent changes in the banana production of Middle America', *Annals of the Association of American Geographers*, vol. 32 (1942), pp. 371–83; C. F. Jones and P. C. Morrison, 'Evolution of the banana industry of Costa Rica', *Economic Geography*, vol. 28 (1952), pp. 1–19; D. A. Preston, 'Changes in the economic geography of banana production in Ecuador', *Transactions and Papers of the*

Institute of British Geographers, vol. 37 (1965), pp. 77–100.

8   D. B. Grigg, *The Agricultural Systems of the World: An Evolutionary Approach* (Cambridge, 1974), pp. 219–26.

9   E. O. Shann, *An Economic History of Australia* (Cambridge, 1948), pp. 78, 120–3, 130–1, 380; A. Marshall, 'The "environment" and Australian wool production: one hundred and fifty years', in J. Andrews (ed.), *Frontiers and Men: A Volume in Memory of Griffith Taylor (1880–1963)* (Melbourne, 1966), pp. 120–37; Grigg, *The Agricultural Systems*, p. 251.

10 M. Capstick, *The Economics of Agriculture* (London, 1970), p. 59; W. N. Peach and J. A. Constantine, *Zimmerman's World Resources and Industries* (New York, 1972), pp. 268–9, 271, 288.

11 H. F. Breimeyer, 'The three economies of agriculture', in American Economic Association, *Readings in the Economics of Agriculture* (London, 1970), p. 20.

12 W. G. T. Packard, *The History of the Fertiliser Industry in Britain*, The Fertiliser Society, Proceedings no. 19 (London, 1952), p. 10.

13 Bairoch, 'Agriculture and the industrial revolution', p. 465; F. Dovring, 'Eighteenth century changes in European agriculture: a comment', *Agricultural History*, vol. 43 (1969), pp. 181–6.

14 G. Thiede, 'Technological advances and growth of food production', *European Review of Agricultural Economics*, vols. 1–2 (1973), p. 218; N. F. Jensen, 'Limits to growth in world food production', *Science*, vol. 201 (1978), pp. 317–20.

15 G. Ordish, *The Great Wine Blight* (London, 1972); M. J. Way, 'Crop losses by insects and the problem of control', *Proceedings of the Nutrition Society*, vol. 20 (1960), pp. 6–8; FAO, *The State of Food and Agriculture 1974* (Rome, 1975), p. 130; S. Hays, *The Chemical and Allied Industries: Studies in the British Economy* (London, 1973), pp. 13, 22; G. E. Petersen, 'The discovery and development of 2,4–D', *Agricultural History*, vol. 41 (1967), pp. 243–53.

16 G. E. Fussell, *Jethro Tull: His Influence on Mechanized Agriculture* (London, 1973), p. 89; Sir John Clapham, *The Economic Development of France and Germany 1815–1914* (Cambridge, 1961), p. 170; R. H. Anderson, 'Grain drills through 39 centuries', *Agricultural History*, vol. 10 (1936), pp. 157–205; F. A. Shannon, *The Farmer's Last Frontier: Agriculture 1860–1897* (New York, 1961), p. 131.

17 H. C. Darby, 'The draining of the English claylands', *Geographische Zeitschrift*, vol. 52 (1964), pp. 190–201; E. L. Jones and E. J. T. Collins, 'Sectoral advance in English agriculture, 1850–1880', *Agricultural History Review*, vol. 15 (1967), pp. 65–81; F. W. H. Green, 'Recent changes in land use and treatment', *Geographical Journal*, vol. 142 (1976), pp. 12–26; A. D. M. Phillips and H. D. Clout, 'Underdraining in France during the second half of the nineteenth century', *Transactions and Papers of the Institute of British Geographers*, vol. 51 (1970), pp. 55–70.

18 F. M. L. Thompson, 'The second agricultural revolution, 1815–1880', *Economic History Review*, vol. 21 (1968–9), pp. 62–77; Packard, *The History of the Fertiliser Industry*, pp. 8–34; D. W. F. Hardie and J. Davidson Pratt, *A History of the Modern British Chemical Industry* (Oxford, 1966), pp. 48, 51–2; L. F. Haber, *The Chemical Industry During the Nineteenth Century: A Study of the Economic Aspect of Applied Chemistry in Europe and North America* (Oxford, 1958), pp. 60–6, 106.

19 W. Rasmussen, 'The impact of technological change in American agriculture, 1862–1892', *Journal of Economic History*, vol. 22 (1962), pp. 578–91; Sir Joseph Hutchinson, *Farming and Food Supply: The Interdependence of Countryside and Town* (Cambridge, 1977), pp. 58–9; J. P. Johnson and D. J. Halliday, 'The development of fertiliser use in the U.K. since 1945', in A. H. Bunting (ed.), *Change in Agriculture* (London, 1970), pp. 265–633; F. Dovring, 'The transformation of European agriculture', in H. J. Habakkuk and M. M. Postan (eds.), *Cambridge Economic History of Europe*; vol. 6, *The Industrial Revolutions and After: Incomes, Population and Technological Change*, part 2 (Cambridge, 1965), pp. 604–69; Z. Griliches, 'Agriculture: productivity and technology', *International Encyclopaedia of Social Science* (New York, 1968), vol. 1, p. 243; FAO, *Production Yearbook, 1962*, vol. 16 (Rome, 1963), pp. 257–63; *Production Yearbook 1976*, vol. 30 (Rome, 1977), pp. 258–77.

20 W. Malenbaum, *The World Wheat Economy 1885–1939* (Cambridge, Mass., 1953), pp. 240–1.

21 B. H. Slicher van Bath, 'Agriculture in the Low Countries', *Relazioni del X Congresso Internazionale di Scienze Storiche, Storia Moderna*, vol. 4 (1955), pp. 169–203; B. H. Slicher van Bath, *The Agrarian History of Western Europe, AD 500–1850* (London, 1963), p. 175; B. H. Slicher van Bath, 'Agriculture in the vital revolution', in C. H. Wilson and E. E. Rich (eds.), *Cambridge Economic History of Europe*; vol. 5, *The Economic Organization of Early Modern Europe* (Cambridge, 1977), p. 98; E. J. T. Collins, 'The diffusion of the threshing machine in Britain, 1790–1880', *Tools and Tillage*, vol. 2 (1972), p. 18; C. Danhof, 'Tools and implements of agriculture', *Agricultural History*, vol. 46 (1972), pp. 81–90; E. D. Ross, 'Retardation in farm technology before the Power Age', *Agricultural History*, vol. 30 (1966), p. 16; Bairoch, 'Agriculture and the industrial revolution', p. 465.

22 G. T. Barton and M. R. Cooper, 'Relations of agricultural production to inputs', *Review of Economics and Statistics*, vol. 30 (1948), pp. 117–26; R. G. Dunbar, 'The role of agricultural history in economic development', *Agricultural History*, vol. 41 (1967), pp. 329–44.

23 E. J. T. Collins, 'Labour supply and demand in European agriculture 1850–1880', in E. L. Jones and S. J. Woolf (eds.), *Agrarian Change and Economic Development: The Historical Problems*, (London,

1969), pp. 61–94; Shannon, *The Farmer's Last Frontier*, p. 144; Rasmussen, 'The impact of technological change', p. 587.

24 E. H. Whetham, 'The mechanization of British farming 1910–1945', *Journal of Agricultural Economics*, vol. 21 (1970), pp. 317–31.

Chapter 10   Transport and agricultural change

1   P. Hall (ed.), *Von Thünen's Isolated State* (Oxford, 1966).

2   A. Lösch, *The Economics of Location* (New Haven, Conn., 1954), pp. 36–59; E. S. Dunn, *The Location of Agricultural Production* (Gainesville, Fla, 1967), pp. 5–12.

3   T. C. Barker and C. I. Savage, *An Economic History of Transport in Britain* (London, 1974), p. 19; J. Crofts, *Packhorse, Waggon and Post: Land Carriage and Communications Under the Tudors and Stuarts* (London, 1967), p. 1; J. A. Chartres, 'Road carrying in England in the seventeenth century: myth and reality', *Economic History Review*, vol. 30 (1977), pp. 73–94; Crofts, *Packhorse, Waggon and Post*, p. 2; D. R. Ringrose, 'Transportation and economic stagnation in eighteenth century Castile', *Journal of Economic History*, vol. 28 (1968), p. 52.

4   E. Pawson, *Transport and Economy: The Turnpike Roads of Eighteenth Century Britain* (London, 1977); M. M. Postan, 'The trade of medieval Europe: the north', in M. M. Postan and E. E. Rich (eds.), *Cambridge Economic History of Europe*; vol. 2, *Trade and Industry in the Middle Ages* (Cambridge, 1952), pp. 151–2; D. Baker, 'The marketing of corn in the first half of the eighteenth century: north east Kent', *Agricultural History Review*, vol. 18 (1970), pp. 126–50.

5   M. Chisholm, *Rural Settlement and Land Use* (London, 1979), pp. 38–40.

6   H. L. Gray, *English Field Systems* (London, 1959), pp. 158–61.

7   G. B. G. Bull, 'Thomas Milne's land utilization map of the London area in 1800', *Geographical Journal*, vol. 122 (1956), pp. 25–30; P. J. Atkins, 'London's intra-urban milk supply, *circa* 1790–1914', *Transactions and Papers of the Institute of British Geographers*, vol. 2, new series (1973), pp. 383–93; F. J. Fisher, 'The development of the London food market, 1540–1640', *Economic History Review*, vol. 5 (1934–5), pp. 46–64; J. Chartres, *Internal Trade in England, 1500–1700* (London, 1977), pp. 19–24.

8   J. T. Yelling, *Common Field and Enclosure in England 1450–1850* (London, 1977), pp. 15, 36–9.

9   W. Abel, *Agricultural Fluctuations in Europe. From the Thirteenth to the Twentieth Centuries* (London, 1980), pp. 106–14.

10 J. H. Parry, 'Transport and trade routes', in E. E. Rich and C. H. Wilson (eds.), *Cambridge Economic History of Europe*; vol. 4, *The Economy of Expanding Europe in the Sixteenth and Seventeenth*

Centuries (Cambridge, 1967), p. 183.

11 J. Nõu, *The Development of Agricultural Economics in Europe* (Uppsala, 1967); E. T. Benedict, H. H. Stoppler and M. R. Benedict (eds.), *Theodore Brinkmann's Economics of the Farm Business* (Berkeley, 1935), pp. 34–49; Dunn, *The Location of Agricultural Production*, pp. 74–9.

12 Pawson, *Transport and Economy*, p. 30; W. Albert, *The Turnpike Road System in England 1663–1840* (Cambridge, 1972), p. 186; P. S. Bagwell, *The Transport Revolution from 1770* (London, 1974), p. 2; D. C. North, 'Sources of productivity change in oceanic shipping, 1600–1850', *Journal of Political Economy*, vol. 76 (1968), pp. 953–70.

13 J. R. Scobie, *Revolution on the Pampas: A Social History of Argentine Wheat 1860–1910* (Austin, Texas, 1967), p. 10; G. R. Hawke, *Railways and Economic Growth in England and Wales 1840–1870* (Oxford, 1970), p. 156; North, 'Sources of productivity change'; D. Howells, 'The impact of railways on agricultural development in nineteenth century Wales', *Welsh Historical Review*, vol. 7 (1974), pp. 40–62; W. Malenbaum, *The World Wheat Economy, 1885–1939*, Harvard Economic Studies no. 92 (Boston, Mass., 1953), p. 39; J. R. Peet, 'The spatial expansion of commercial agriculture in the nineteenth century: a von Thünen interpretation', *Economic Geography*, vol. 45 (1969), p. 296; D. C. North, *Growth and Welfare in the American Past: A New Economic History*, (Englewood Cliffs, NJ, 1974), p. 113; D. C. North, 'Ocean freight rates and economic development, 1750–1913', *Journal of Economic History*, vol. 18 (1958), p. 545; M. Falkus, 'Russia and the international wheat trade', *Economica*, vol. 33 (1966), pp. 416–29.

14 T. N. Morris, 'Management and preservation of food', in C. Singer, E. J. Holmyard, A. R. Hall and T. I. Williams (eds.), *A History of Technology*, vol. 5 (Oxford, 1958), pp. 16–48; A. J. Youngson, 'The opening up of new territories', in H. J. Habakkuk and M. M. Postan (eds.), *The Cambridge Economic History of Europe*; vol. 6, *The Industrial Revolutions and After*, pt. 2 (Cambridge, 1965), p. 172.

15 P. J. Atkins, 'The growth of London's railway milk trade *c.* 1845–1914', *Journal of Transport History*, vol. 4 (1977–8), pp. 208–26; E. S. Simpson, 'Milk production in England and Wales: a study in the influence of collective marketing', *Geographical Review*, vol. 49 (1959), pp. 95–111; E. H. Whetham, 'The London milk trade, 1860–1900', *Economic History Review*, vol. 17 (1964), pp. 369–80; F. A. Barnes, 'The evolution of the salient patterns of milk production and distribution in England and Wales', *Transactions and Papers of the Institute of British Geographers*, vol. 25 (1958), pp. 167–95; D. Grigg, *The Agricultural Systems of the World: An Evolutionary Approach* (Cambridge, 1974), p. 195.

16 J. R. Peet, 'Von Thünen theory and the dynamics of agricultural

expansion', *Explorations in Economic History*, vol. 8 (1970), pp. 180–201.

17   R. Abler, J. S. Adams and P. Gould, *Spatial Organization: The Geographer's View of the World* (Englewood Cliffs, NJ, 1971), pp. 351–2.

18   United States Department of Agriculture, *Yearbook, 1921* (Washington DC, 1922), p. 90.

19   J. E. Spencer and R. J. Horvath, 'How does an agricultural region originate?', *Annals of the Association of American Geographers*, vol. 53 (1963), pp. 74–92.

20   C. W. Larson, 'The dairy industry', in United States Department of Agriculture, *Yearbook, 1922* (Washington DC, 1923), pp. 298–314; L. Durand, 'The migration of cheese manufacturing in the United States', *Annals of the Association of American Geographers*, vol. 42 (1952), pp. 263–82.

21   J. T. Schlebecker, 'The world metropolis and the history of American agriculture', *Journal of Economic History*, vol. 20 (1960), pp. 187–208; Peet, 'The spatial expansion', pp. 283–301; J. R. Peet, 'The influences of the British market on agriculture and related economic development before 1860', *Transactions and Papers of the Institute of British Geographers*, vol. 56 (1972), pp. 1–20.

22   C. T. Smith, *An Historical Geography of Western Europe before 1800* (London, 1967), p. 497.

Chapter 11   The diffusion of agricultural innovations

1   R. Heine Geldern, 'Cultural diffusion', *International Encyclopedia of the Social Sciences*, vol. 4 (New York, 1968), pp. 169–73; E. Katz, M. Levin and H. Hamilton, 'Traditions of research on the diffusion of information', *American Sociological Review*, vol. 28 (1963), pp. 237–52; L. A. Brown and E. G. Moore, 'Diffusion research in geography: a perspective', *Progress in Geography*, vol. 1 (1969), pp. 121–57.

2   E. Whetham, 'The mechanization of British farming 1910–1945', *Journal of Agricultural Economics*, vol. 21 (1970), pp. 317–31; E. J. T. Collins, 'The diffusion of the threshing machine in Britain, 1790–1880', *Tools and Tillage*, vol. 2 (1972), pp. 16–33.

3   G. E. Jones, 'The adoption and diffusion of agricultural practices', *World Agricultural Economics and Rural Sociology Abstracts*, vol. 9 (1967), p. 15.

4   F. C. Fliegel and J. E. Kivlin, 'Farm practice attributes and adoption rates', *Social Forces*, vol. 40 (1962), pp. 364–70.

5   G. E. Jones, 'The diffusion of agricultural innovations', in I. Burton and R. W. Kates (eds.), *Readings in Resource Management and Conservation* (Chicago, 1965), pp. 475–92.

6 P. Gould, *Spatial Diffusion* (Washington DC, 1969); T. Hägerstrand, *Innovation Diffusion as a Spatial Process* (Chicago, 1967).

7 M. Overton, 'Computer analysis of an incomplete data source: the case of the probate inventories', *Journal of Historical Geography*, vol. 3 (1977), pp. 317–26; A. Harris, *The Rural Landscape of the East Riding of Yorkshire 1700–1850: A Study in Historical Geography* (London, 1961), p. 86.

8 R. H. Tawney, *The Agrarian Problem in the Sixteenth Century* (London, 1912); H. J. Habakkuk, 'English landownership 1680–1740', *Economic History Review*, vol. 10 (1940), pp. 2–17; Lord Ernle, *English Farming Past and Present* (London, 1961).

9 G. Berg, 'The introduction of the winnowing machine in Europe in the eighteenth century', *Tools and Tillage*, vol. 3 (1976), pp. 24–46; J. Kuise, 'Mechanisation, commercialisation and the' protectionist movement in Swedish agriculture, 1860–1910', *Scandinavian Economic Review*, vol. 19 (1971), pp. 23–4; J. R. Walton, 'Mechanization in agriculture: a study of the adoption process', in H. S. Fox and R. A. Butlin (eds.), *Change in the Countryside: Essays on Rural England 1500–1900*, Institute of British Geographers Special Publication no. 10 (London, 1979), pp. 23–42; A. L. Olmstead, 'The mechanization of reaping and mowing in American agriculture, 1833–1870', *Journal of Economic History*, vol. 35 (1975), pp. 327–52.

10 B. H. Slicher van Bath, 'The rise of intensive husbandry in the Low Countries', in J. S. Bromley and E. H. Kossmann (eds.), *Britain and the Netherlands*, vol. 1 (London, 1960), p. 148; D. B. Grigg, 'The development of tenant right in South Lincolnshire', *Lincolnshire Historian*, vol. 2 (1962), pp. 41–8.

11 G. E. Fussell, *The Old English Farming Books from Fitzherbert to Tull 1523–1730* (London, 1947); G. E. Fussell, *More Old English Farming Books from Tull to the Board of Agriculture, 1731 to 1793* (London, 1950); H. S. Fox, 'Local farmers' associations and the circulation of agricultural information in nineteenth century England', in Fox and Butlin (eds.), *Change in the Countryside*, pp. 43–63; A. J. Bourde, *The Influence of England on the French Agronomes 1750–1789* (Cambridge, 1953).

12 B. H. Slicher van Bath, 'Agriculture in the vital revolution', in C. H. Wilson and E. E. Rich (eds.), *Cambridge Economic History of Europe*; vol. 4, *The Economic Organization of Early Modern Europe* (Cambridge, 1977), p. 99; C. M. Cipolla, *Literacy and Development in the West* (London, 1969).

13 Fox, 'Local farmers' associations', pp. 43–63; S. McDonald, 'The diffusion of knowledge among Northumberland farmers 1780–1815', *Agricultural History Review*, vol. 27 (1979), pp. 30–9; A. Fenton, 'Sickle, scythe and reaping machine: innovation patterns in Scotland', *Ethnologaea*, vol. 7 (1973–4), pp. 35–47.

14   A. Findley and D. Maclennan, 'Innovation diffusion at the micro-scale', *Area*, vol. 10 (1978), pp. 309–14.

15   J. G. D. Clark, 'Radio carbon dating and the expansion of farming culture from the Near East over Europe', *Proceedings of the Prehistoric Society*, vol. 31 (1965), pp. 58–77; C. W. Bishop, 'Origin and early diffusion of the traction-plough', *Antiquity*, vol. 10 (1936), pp. 261–81; Z. Griliches, 'Hybrid corn and the economics of innovation', *Science*, vol. 132 (1960), pp. 275–80; L. Seig, 'The spread of tobacco: a study in cultural diffusion', *Professional Geographer*, vol. 15 (1963), pp. 17–20.

16   A. J. Ammerman and L. L. Cavalli-Sforza, 'Measuring the rate of spread of early farming in Europe', *Man*, vol. 6 (1971), pp. 674–88.

17   Jones, 'The diffusion of agricultural innovations', pp. 475–92.

Chapter 12   The definition and measurement of production and productivity growth in agriculture

1   B. F. Johnston and J. T. Nielsen, 'Agriculture and structural transformation in a developing economy', *Economic Development and Cultural Change*, vol. 14 (1966), pp. 279–301.

2   F. Dovring, 'Eighteenth century changes in European agriculture: a comment', *Agricultural History*, vol. 43 (1969), pp. 181–6.

3   Paul Bairoch, 'Agriculture and the industrial revolution, 1700–1914', in C. M. Cipolla (ed.), *The Fontana Economic History of Europe*; vol. 3, *The Industrial Revolution* (London, 1973), pp. 457–60.

4   D. S. Landes, *The Unbound Prometheus: Technological Change and Industrial Development in Western Europe from 1750 to the Present* (London, 1970), p. 41.

5   Paul Bairoch, 'Niveaux de développement économique de 1810 à 1910', *Annales ESC*, vol. 20 (1965), pp. 1091–117; Paul Bairoch, *The Economic Development of the Third World since 1900* (London, 1975); A. Daniel, 'Regional differences of productivity in European agriculture', *Review of Economic Studies*, vol. 12 (1944–5), pp. 50–70.

6   C. J. Doyle, 'A comparative study of agricultural productivity in the UK and Europe', *Journal of Agricultural Economics*, vol. 30 (1979), pp. 261–70.

7   P. Deane and W. A. Cole, *British Economic Growth, 1688–1959: Trends and Structure* (Cambridge, 1962), pp. 166, 170; M. Towne and W. D. Rasmussen, 'Farm gross product and gross investment in the nineteenth century', in National Bureau of Economic Research, *Trends in the American Economy in the Nineteenth Century*, Studies in Income and Wealth, vol. 24 (Princeton, NJ, 1960), pp. 257–60.

8   FAO, *World Agriculture: The Last Quarter Century* (Rome, 1970), p. 9; FAO, *Production Yearbooks* (Rome, 1961 *et seq.*).

9   J. C. Toutain, *Le produit de l'agriculture français* (Paris, 1961), vol. 1, pp. 213–15; Deane and Cole, *British Economic Growth*, pp. 65, 78.

10   E. Le Roy Ladurie, 'De la crise ultime à la vraie croissance', in E. le Roy Ladurie (ed.), *Histoire de la France rurale*; vol. 2, *L'âge classique des paysans, 1340–1789* (Paris, 1975), p. 582.

11   Y. Hayami and V. W. Ruttan, *Agricultural Development: An International Perspective* (London, 1971), pp. 327–31.

12   J. Lingard and A. J. Rayner, 'Productivity growth in agriculture: a measurement framework', *Journal of Agricultural Economics*, vol. 25 (1975), pp. 87–103.

13   R. E. Gallman, 'Changes in total US factor productivity growth in the nineteenth century', *Agricultural History*, vol. 46 (1972), pp. 191–210; Anon., 'Productivity measurement in agriculture', *Economic Trends*, no. 91 (1961), pp. ii–v.

14   Deane and Cole, *British Economic Growth*, p. 75; Toutain, *Le produit de l'agriculture français*; Ladurie, 'De la crise ultime', p. 582.

15   P. K. O'Brien, D. Heath and C. Keyder, 'Agriculture in Britain and France, 1815–1914', *Journal of European Economic History*, vol. 6 (1977), p. 341; Toutain, *Le produit de l'agriculture français*.

16   J. Kendrick, *Post War Productivity Trends in the United States, 1948–1969* (New York, 1973).

17   Z. Griliches, 'Measuring inputs in agriculture: a critical survey', *Journal of Farm Economics*, vol. 42 (1960), pp. 1411–27.

18   C. A. Chandler, 'The relative contribution of capital intensity and productivity to changes in output and income in the U.S. economy, farm and non-farm sectors, 1946–1958', *Journal of Farm Economics*, vol. 44 (1962), pp. 335–48.

19   R. A. Berry and W. R. Cline, *Agrarian Structure and Productivity in Developing Countries* (Baltimore, Md, 1979).

Chapter 13   Agricultural revolutions in Europe

1   G. Duby, *Rural Economy and Country Life in the Medieval West* (London, 1962); G. Duby, 'La révolution agricole médiévale', *Revue de Géographie du Lyon*, vol. 29 (1954), pp. 361–6; G. Duby, 'Medieval agriculture 900–1500', in C. M. Cipolla (ed.), *The Fontana Economic History of Europe*; vol. 1, *The Middle Ages* (London, 1972), pp. 175–220; Lynn T. White Jr, *Medieval Technology and Social Change* (Oxford, 1962); Lynn T. White Jr, 'The expansion of technology, 500–1500', in Cipolla (ed.), *The Middle Ages*, pp. 143–75.

2   J. Thirsk, 'The common fields', *Past and Present*, vol. 29 (1964), pp. 3–25; R. C. Hoffman, 'Medieval origins of the common fields', in W. N. Parker and E. L. Jones (eds.), *European Peasants and their Markets* (Princeton, NJ, 1975), pp. 23–72.

3   White, *Medieval Technology*, pp. 40–78; Duby, 'La révolution', pp.

361–6; J. Z. Titow, *English Rural Society 1200–1350* (London, 1969), pp. 39–40; G. Fourquin, 'Paysannerie et féodalité', in G. Duby and A. Wallon (eds.), *Histoire de la France rurale*; vol. 1, *L'âge classique des paysans 1340–1789* (Paris, 1975), pp. 406–12.

4   G. E. Fussell, 'Ploughs and ploughing before 1800', *Agricultural History*, vol. 40 (1966), pp. 177–86; B. Wailes, 'Plow and population in temperate Europe', in B. Spooner (ed.), *Population Growth: Anthropological Implications* (Cambridge, Mass., 1972), pp. 154–79.

5   Duby, 'Medieval agriculture', p. 186; E. M. Jope, 'Agricultural implements', in C. Singer, E. J. Holmyard, A. R. Hall and T. I. Williams (eds.), *A History of Technology*, vol. 4 (Oxford, 1957), p. 94.

6   J. Blum, *The End of the Old Order in Rural Europe* (Princeton, NJ, 1978), p. 192.

7   A. Smith, 'Regional differences in crop production in medieval Kent', *Archaeologia Cantiana*, vol. 28 (1963), pp. 147–64; J. Z. Titow, *Winchester Yields: A Study in Medieval Agricultural Productivity* (Cambridge, 1972), p. 31; P. F. Brandon, 'Arable farming in a Sussex scarp foot parish during the late Middle Ages', *Sussex Archaeological Collections*, vol. 100 (1962), pp. 60–72; Duby, *Rural Economy*, pp. 97, 104; Fourquin, 'Paysannerie et féodalité', p. 418.

8   Duby, 'Medieval agriculture', p. 196; Titow, *Winchester Yields*, p. 30; M. M. Postan, 'Medieval agrarian society in its prime: England', in M. M. Postan (ed.), *Cambridge Economic History of Europe*; vol. 1, *The Agrarian Life of the Middle Ages* (Cambridge, 1966), pp. 556–8; D. Roden, 'Demesne farming in the Chiltern Hills', *Agricultural History*, vol. 17 (1969), pp. 9–23; P. F. Brandon, 'Demesne arable farming in coastal Sussex during the later Middle Ages', *Agricultural History Review*, vol. 19 (1971), pp. 113–34.

9   Duby, 'Medieval agriculture', pp. 25–6; B. H. Slicher van Bath, 'The yields of different crops (mainly cereals) in relation to the seed *c.* 810–1820', *Acta Historiae Neerlandica*, vol. 2 (1967), pp. 32–64.

10   K. Marx, *Capital*, vol. 1 (London, 1977), p. 697; A. Toynbee, *The Industrial Revolution* (Boston, Mass., 1956), pp. 12–18; Lord Ernle, *English Farming: Past and Present* (London, 1961), pp. 148–252.

11   Ernle, *English Farming*, p. 148.

12   M. A. Havinden, 'Agricultural progress in open-field Oxfordshire', *Agricultural History Review*, vol. 9 (1961), pp. 73–83; T. H. Marshall, 'Jethro Tull and the new husbandry of the eighteenth century', *Economic History Review*, vol. 2 (1929), pp. 41–60; M. Turner, *English Parliamentary Enclosure: Its Historical Geography and Economic History* (Folkestone, 1980), pp. 32–62.

13   R. W. Sturgess, 'The agricultural revolution on the English clays', *Agricultural History Review*, vol. 14 (1966), pp. 104–21; E. J. T. Collins and E. L. Jones, 'Sectoral advance in English agriculture, 1850–1880', *Agricultural History Review*, vol. 15 (1967), pp. 65–81.

14 R. A. C. Parker, *Coke of Norfolk: A Financial and Agricultural Study, 1707–1842* (Oxford, 1975), pp. 106, 154.

15 E. Kerridge, *The Agricultural Revolution* (London, 1967).

16 E. Kerridge, 'The agricultural revolution reconsidered', *Agricultural History*, vol. 43 (1969), p. 464–74; G. E. Mingay, 'The agricultural revolution in English history: a reconsideration', *Agricultural History*, vol. 26 (1963), pp. 123–33; G. E. Mingay, 'Dr. Kerridge's *Agricultural Revolution*: a comment', *Agricultural History*, vol. 43 (1969), pp. 477–81; G. E. Mingay, *The Agricultural Revolution: Changes in Agriculture, 1650–1880* (London, 1977); E. L. Jones, 'Introduction', in E. L. Jones (ed.), *Agriculture and Economic Growth, 1650–1815* (London, 167), pp. 1–48.

17 F. M. L. Thompson, 'The second agricultural revolution, 1815–1880', *Economic History Review*, vol. 21 (1968), pp. 62–77.

18 P. Deane and W. A. Cole, *British Economic Growth 1688–1959: Trends and Structure* (Cambridge, 1962), pp. 62–74, 152, 155, 158, 170; L. Drescher, 'The development of agricultural production in Great Britain and Ireland from the early nineteenth century', *Manchester School of Economic and Social Studies,* vol. 23 (1955), pp. 153–75; E. M. Ojala, *Agriculture and Economic Progress* (London, 1952), pp. 210–15.

19 Drescher, 'The development of agricultural production', p. 167.

20 G. E. Fussell (ed.), *Robert Loder's Farm Accounts 1610–1620*, Camden 3rd Series, vol. 53 (London, 1936).

21 Assuming a sown area of 2.83 million hectares in 1700 (Table 27) and a yield of 1009 kg/ha; in 1850 the sown area would have been about 5.46 million hectares and the average yield 1781 kg/ha. The cereal area in 1700 was probably of the order of 2.02 million hectares and 3.1 million hectares in the 1850s.

22 J. D. Chambers, 'Enclosure and labour supply in the industrial revolution', *Economic History Review*, vol. 5 (1952–3), pp. 319–43; C. P. Timmer, 'The turnip, the new husbandry and the English agricultural revolution', *Quarterly Journal of Economics*, vol. 83 (1969), pp. 375–95.

23 P. K. O'Brien, 'Agriculture and the industrial revolution', *Economic History Review*, vol. 30 (1977), pp. 166 .81; P. H. Lindert, 'English occupations, 1670–1811', *Journal of Economic History*, vol. 40 (1980), pp. 685–712.

24 B. Holderness, 'Capital formation in agriculture', in J. P. Higgins and S. Pollard (eds.), *Aspects of Capital Formation in Great Britain, 1750–1850: A Preliminary Survey* (London, 1975), pp. 159–83.

25 E. H. Hunt, 'Labour productivity in English agriculture, 1850–1914', *Economic History Review*, vol. 20 (1967), pp. 280–302.

26 O'Brien, 'Agriculture and the industrial revolution', p. 169.

Chapter 14   Class, region and revolution

1   D. C. North and R. P. Thomas, *The Rise of the Western World: A New Economic History* (Cambridge, 1973); E. T. Thompson, 'Population expansion and the plantation system', *American Journal of Sociology*, vol. 41 (1935), pp. 30–45.

2   Karl Marx, *Capital*, vol. 1 (London, 1977), pp. 667–712; Karl Marx, *Grundrisse: Foundations of the Critique of Political Economy* (London, 1973).

3   M. Dobb, *Studies in the Development of Capitalism* (London, 1963); P. Anderson, *Passages from Antiquity to Feudalism* (London, 1974), pp. 147–97; R. Brenner, 'Agrarian class structure and economic development in pre-industrial Europe', *Past and Present*, no. 70 (1976), pp. 30–75; R. H. Hilton, 'A "crisis of feudalism"', *Past and Present*, no. 80 (1978), pp. 3–19; Dobb, *Studies*, pp. 20, 42, 52, 63–5; J. S. Cohen, 'The achievements of economic history: the Marxist school', *Journal of Economic History*, vol. 38 (1978), pp. 33–40; R. H. Hilton (ed.), *The Transition from Feudalism to Capitalism* (London, 1976); Marx, *Capital*, 1977 vol. 1, pp. 669–72; E. J. Nell, 'Economic relationships in the decline of feudalism: an examination of economic interdependence and social change', *History and Theory*, vol. 6 (1967), pp. 313–50.

4   Marx, *Capital*, vol. 1, pp. 668, 669, 671, 672, 675, 676, 678, 681, 694.

5   ibid., p. 651.

6   H. J. Habakkuk, 'English landownership 1680–1740', *Economic History Review*, vol. 10 (1939–40), pp. 2–17. J. V. Beckett, 'English landownership in the later seventeenth and eighteenth centuries: the debate and problems', *Economic History Review*, vol. 30 (1977), pp. 567–81.

7   J. T. Yelling, *Common Field and Enclosure in England 1450–1850* (London, 1977), p. 21; R. Butlin, 'The enclosure of open fields and extinction of common rights in England, *circa* 1600–1750: a review', in H. S. Fox and R. Butlin (eds.), *Change in the Countryside: Essays on Rural England 1500–1900*, Institute of British Geographers Special Publication no. 10 (London, 1979), pp. 65–82.

8   R. H. Tawney *The Agrarian Problem in the Sixteenth Century* (London, 1912).

9   E. F. Gay, 'Inclosures in England in the sixteenth century', *Quarterly Journal of Economics*, vol. 17 (1903), pp. 576–97.

10 M. Beresford, 'A review of historical research (to 1968)', in M. Beresford and J. G. Hurst (eds.), *Deserted Medieval Villages* (London, 1971), pp. 11–12, 16–17.

11 E. C. K. Gonner, *Common Land and Inclosure* (London, 1912).

12 G. Slater, *The English Peasantry and the Enclosure of Common Fields* (1909); Gonner, *Common Land*.

13 W. E. Tate, *A Domesday of English Enclosure Acts and Awards*, ed.

M. E. Turner, (Reading, 1978); M. E. Turner, *English Parliamentary Enclosure: Its Historical Geography and Economic History* (Folkestone, 1980).

14 Turner, *English Parliamentary Enclosure*, p. 24.

15 Marx, *Capital*, vol. 1, p. 669.

16 ibid., pp. 671, 676.

17 J. P. Cooper, 'The social distribution of land and men in England, 1436–1700', *Economic History Review*, vol. 20 (1967), pp. 419–40; F. M. L. Thompson, 'The social distribution of landed property in England since the sixteenth century', *Economic History Review*, vol. 19 (1966), pp. 505–17.

18 M. E. Turner, 'Parliamentary enclosure and landownership change in Buckinghamshire', *Economic History Review*, vol. 28 (1975), pp. 565–81; G. E. Mingay, *Enclosure and the Small Farmer in the Age of the Industrial Revolution* (London, 1968).

19 Marx, *Capital*, vol. 1, p. 671; V. I. Lenin, *Collected Works*; vol. 3, *The Development of Capitalism in Russia* (London, 1960), pp. 70–140; R. H. Hilton, 'Reasons for inequality among medieval peasants', *Journal of Peasant Studies*, vol. 5 (1977–8), pp. 271–84.

20 M. Spufford, *Contrasting Communities: English Villages in the Sixteenth and Seventeenth Centuries* (Cambridge, 1974), pp. 46–51, 65–70, 118, 165.

21 J. L. and B. Hammond, *The Village Labourer, 1760–1832: A Study in the Government of England before the Reform Bill* (London, 1911); H. Levy, *Large and Small Holdings* (Cambridge, 1911); G. E. Mingay, 'The size of farms in the eighteenth century', *Economic History Review*, vol. 14 (1961–2), pp. 469–88; J. R. Wordie, 'Social change on the Leveson Gower estates 1714–1832', *Economic History Review*, vol. 27 (1974–5), pp. 593–609.

22 D. Grigg, 'Large and small farms in England and Wales; their size and distribution', *Geography*, vol. 48 (1963), pp. 268–79.

23 Marx, *Capital*, vol. 1, pp. 647, 678, 700; Dobb, *Studies*, pp. 227–51; John Saville, *Marxism and History* (Hull 1974); John Saville, 'Primitive accumulation and early industrialization in Britain', in R. Miliband and J. Saville (eds.), *The Socialist Register 1969* (London, 1969), pp. 247–71.

24 D. Hey, *An English Rural Community: Myddle under the Tudors and Stuarts* (Leicester, 1974), p. 53; V. Skipp, 'Economic and social change in the Forest of Arden, 1530–1649', *Agricultural History Review*, vol. 18 (1970), supplement, pp. 89, 98; A. Everitt, 'Farm labourers', in J. Thirsk, *The Agrarian History of England and Wales*; vol. iv, *1500–1650* (Cambridge, 1967), pp. 398–400.

25 J. D. Chambers, 'Enclosure and labour supply in the Industrial Revolution', *Economic History Review*, vol. 5 (1953), pp. 319–40.

26 Turner, *English Parliamentary Enclosure*, p. 76.

27 Chambers, 'Enclosure and labour supply'; P. Timmer, 'The turnip, the new husbandry and the English agricultural revolution', *Quarterly Journal of Economics*, vol. 83 (1969), pp. 375–95.

28 Mingay, *Enclosure and the Small Farmer*; Wordie, 'Social change'.

29 M. Blaug, 'The myth of the old poor law and the making of the new', *Journal of Economic History*, vol. 23 (1963), pp. 151–84; E. L. Jones, 'The agricultural labour market in England 1793–1872', *Economic History Review*, vol. 17 (1964–5), pp. 322–38.

30 D. Grigg, 'E. G. Ravenstein and the laws of migration', *Journal of Historical Geography*, vol. 3 (1977), pp. 41–54.

31 C. M. Law, 'Some notes on the urban population of England and Wales in the eighteenth century', *Local Historian*, vol. 10 (1972), pp. 13–26; C. M. Law, 'The growth of urban population in England and Wales, 1801–1911', *Transactions and Papers of the Institute of British Geographers*, vol. 41 (1967), pp. 125–44.

Chapter 15   On the frontier

1 F. J. Turner, 'The significance of the frontier in American history', in F. J. Turner, *The Frontier in American History* (New York, 1962), pp. 1–38.

2 R. Hofstadter, *The Progressive Historians: Turner, Beard, Parrington* (London, 1969), p. 94.

3 Turner, *The Frontier in American History*, p. 12.

4 L. M. Hacker, 'Sections – or classes?', in G. R. Taylor (ed.), *The Turner Thesis Concerning the Rôle of the Frontier in American History* (Lexington, Mass., 1972), pp. 51–5; G. W. Pierson, 'The frontier and American institutions: a criticism of the Turner theory', in Taylor (ed.), *The Turner Thesis*, pp. 70–97.

5 R. E. Riegel, 'American frontier theory', *Journal of World History*, vol. 3 (1956), pp. 356–78.

6 T. P. Abernathy, 'The southern frontier: an interpretation', in W. D. Wyman and C. B. Kroeber (eds.), *The Frontier in Perspective* (Madison, Wis., 1957), pp. 129–42; R. A. Billington, *America's Frontier Heritage* (New York, 1966), pp. 39, 45; R. A. Eigenheer, 'The frontier hypothesis and related spatial concepts', *Californian Geographer*, vol. 14 (1973–4), pp. 55–69; H. Bolton, 'The mission as a frontier institution in the Spanish American colonies', *American Historical Review*, vol. 23 (1917), pp. 42–61; G. Blackburn and S. L. Richards, Jr., 'A demographic history of the West: Manistee County, Michigan 1860', *Journal of American History*, vol. 57 (1970), pp. 600–18; J. F. McDermott, 'The frontier re-examined', in J. F. McDermott (ed.), *The Frontier Re-examined* (London, 1967), pp. 1–14.

7 P. W. Gates, 'Frontier estate builders and farm labourers', in Wyman

and Kroeber, *The Frontier in Perspective*, pp. 143–65; P. W. Gates, 'The homestead law in an incongruous land system', *American Historical Review*, vol. 41 (1936), pp. 652–81; Billington, *America's Frontier Heritage*, p. 227.

8 Gates, 'Frontier estate builders'; R. D. Mitchell, 'The commercial nature of frontier settlement in the Shenandoah valley of Virginia', *Proceedings of the Association of American Geographers*, vol. 1 (1969), pp. 109–13.

9 Mitchell, 'The commercial nature'; B. F. Johnston and P. Kilby, *Agriculture and Structural Transformation: Economic Strategies in Late-developing Countries* (Oxford, 1975), p. 199.

10 Billington, *America's Frontier Heritage*, p. 174; W. P. Webb, *The Great Plains* (New York, 1931).

11 Billington, *America's Frontier Heritage*, p. 33.

12 F. A. Shannon, 'A post-mortem on the labour safety-valve theory', *Agricultural History*, vol. 7 (1945), pp. 31–7; C. Goodrich and S. Davison, 'The wage earner in the westward movement, part I', *Political Science Quarterly*, vol. 50 (1935), pp. 161–85, and 'Part II', vol. 51 (1936), pp. 61–116.

13 C. A. Danhof, 'Economic validity of the safety-valve doctrine', *Journal of Economic History*, vol. 1 (1941), pp. 96–106.

14 A. L. Burt, 'If Turner had looked at Canada, Australia and New Zealand when he wrote about the West', in Wyman and Kroeber, *The Frontier in Perspective*, pp. 59–77; D. Gerhard, 'The frontier in comparative view', *Comparative Studies in Society and History*, vol. 1 (1959), pp. 205–29.

15 D. W. Treadgold, *The Great Siberian Migration: Government and Peasant in Resettlement from Emancipation to the First World War* (Princeton, NJ, 1957).

16 H. C. Allen, 'F. J. Turner and the frontier in American history', in H. C. Allen and C. P. Hill (eds.), *British Essays on American History* (London, 1957), pp. 145–60.

17 D. B. Grigg, *The Agricultural Systems of the World: An Evolutionary Approach* (Cambridge, 1974), p. 261.

18 B. Fitzpatrick, 'The Big Man's frontier and Australian farming', *Agricultural History*, vol. 21 (1947), pp. 8–12; A. Hennessy, *The Frontier in Latin American History* (London, 1978), p. 29.

19 Gerhard, 'The frontier in comparative view', p. 213.

Index